Netter's Atlas of the Human Body

Textual material licensed from Harvard Medical School's
Consumer Newsletters and Special Health Reports and
Captions by the Icon Learning Systems' editorial team

Illustrations by
Frank H. Netter, MD

Contributing Illustrators

Carlos A.G. Machado, MD

Kip Carter, MS

John A. Craig, MD

Dragonfly Media Group

James A. Perkins, MS, MFA

BARRON'S

First edition for the United States and Canada published 2006 by Barron's Educational Series, Inc.

Copyright © 2006, prepared for publication, by Icon Learning Systems LLC.

FIRST EDITION

All inquiries should be addressed to:
Barron's Educational Series, Inc.
250 Wireless Boulevard
Hauppauge, New York 11788
http://www.barronseduc.com

Requests for permission should be addressed to Permissions Editor, Icon Learning Systems, www.netterart.com/request.cfm.

ISBN-13: 978-0-7641-5884-1
ISBN-10: 0-7641-5884-8

Library of Congress Catalog No.: 2005920911

NOTICE
Every effort has been made to confirm the accuracy of the information presented and to describe generally accepted practices. Neither the publisher nor the authors can be held responsible for errors or for any conse-quences arising from the use of the information contained herein, and make no warranty, expressed or implied, with respect to the contents of the publication.

Physician Consultant: John A. Craig, MD
Art Director: Jonathan Dimes
Graphic Designers: Colleen Quinn and Nancy Walker
Senior Production Editor: Melanie Peirson Johnstone
Copy Editor: Stephanie Klein
Managing Editor: Alison Hankey
Text Design: Colleen Quinn

Composition and Layout by Icon Learning Systems

Printed in China

10   9   8   7   6   5   4   3   2   1

Table of Contents

Acknowledgements

Special thanks to the following individuals for their unending patience and dedication during the development of this book:

John A. Craig, M.D., physician, artist, and long time contributor to the Netter Collection, who, along with Jonathan Dimes, Art Director for Icon Learning Systems, reviewed our extensive medical art collection and selected the artwork best suited to the topics chosen for this work. Through their efforts, the clearest and most relevant pieces of artwork were chosen and combined to create entirely new plate compositions of which this work is comprised.

Colleen Quinn, Graphic Designer, Icon Learning Systems, for a beautiful internal book design and her Herculean efforts, along with Nancy Walker, in organizing the textual material around the artwork.

Melanie Peirson Johnstone, Senior Production Editor, Icon Learning Systems, for overseeing the final production process and compilation of the index.

Nancy Ferrari, Managing Editor, and Christine Junge, Assistant Editor, Harvard Health Publications, for compiling the textual material from Harvard Medical School's vast database of consumer newsletters and Special Health Reports.

Stephanie Klein, "Editor at Large," for her meticulous editorial, copyediting, and developmental support.

Netter's Atlas of the Human Body

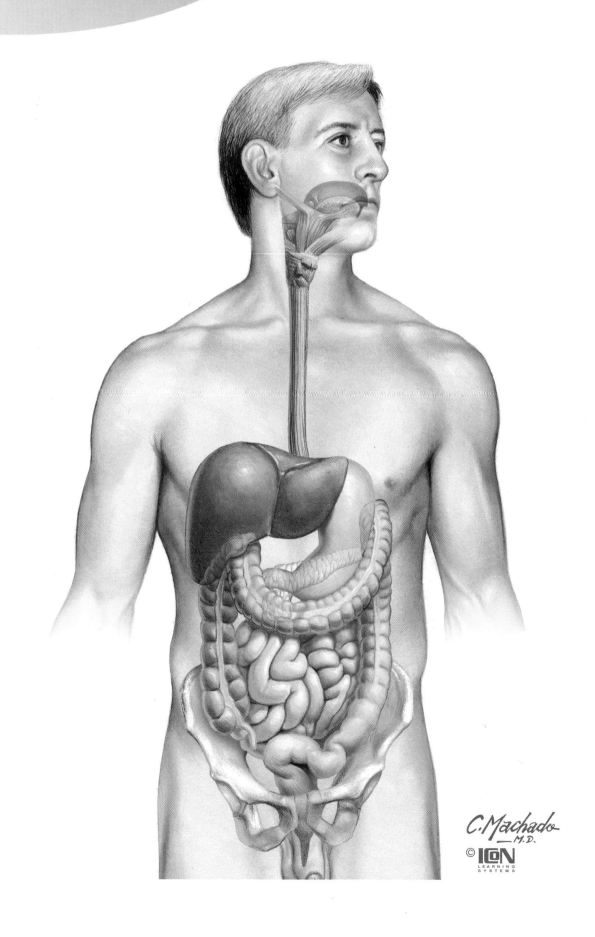

C. Machado
M.D.

© ICON
LEARNING
SYSTEMS

Surface Anatomy

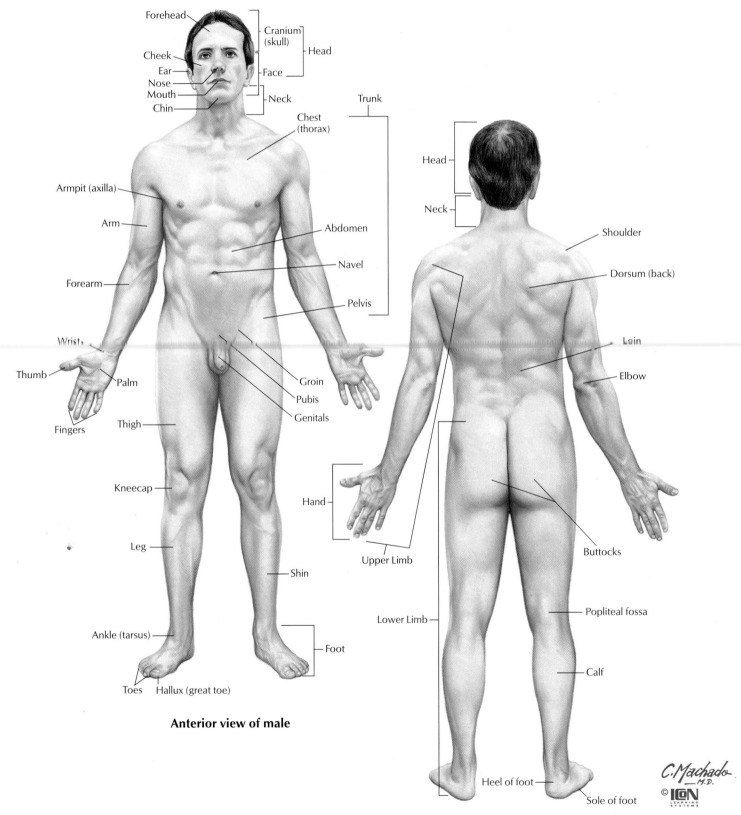

Forehead
Cheek
Ear
Nose
Mouth
Chin
Cranium (skull)
Head
Face
Neck
Trunk
Chest (thorax)
Armpit (axilla)
Arm
Forearm
Abdomen
Navel
Pelvis
Wrist
Thumb
Palm
Fingers
Thigh
Groin
Pubis
Genitals
Kneecap
Hand
Leg
Shin
Ankle (tarsus)
Upper Limb
Foot
Toes   Hallux (great toe)

Anterior view of male

Head
Neck
Shoulder
Dorsum (back)
Loin
Elbow
Buttocks
Lower Limb
Popliteal fossa
Calf
Heel of foot
Sole of foot

Posterior view of male

C. Machado M.D.

Surface features and the descriptive regions of the surface anatomy are best visualized with a person standing in "anatomical position," which is described as

· Standing erect and facing forward

· Arms hanging at the sides with the palms facing forward

· Legs placed together with the feet directed forward

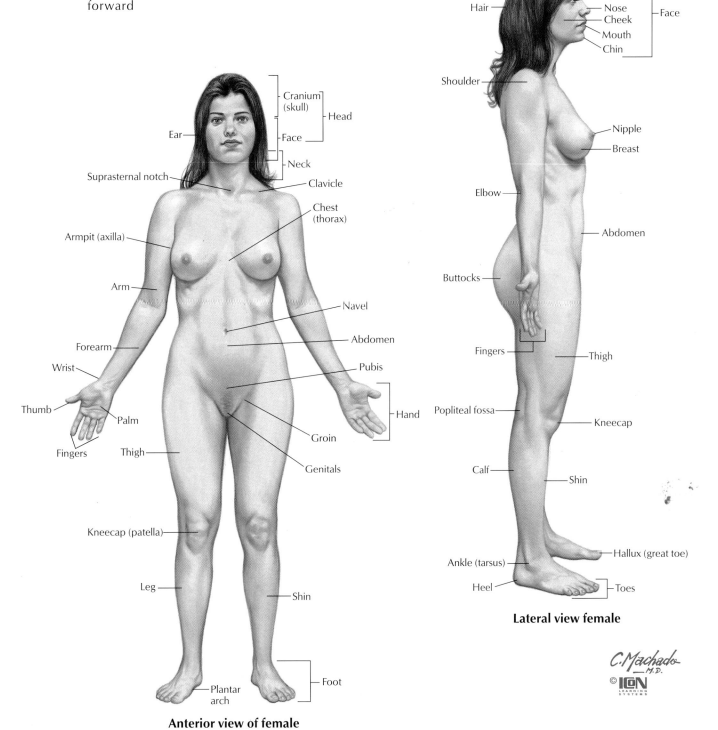

Anterior view of female

Lateral view female

Surface Anatomy
Regions

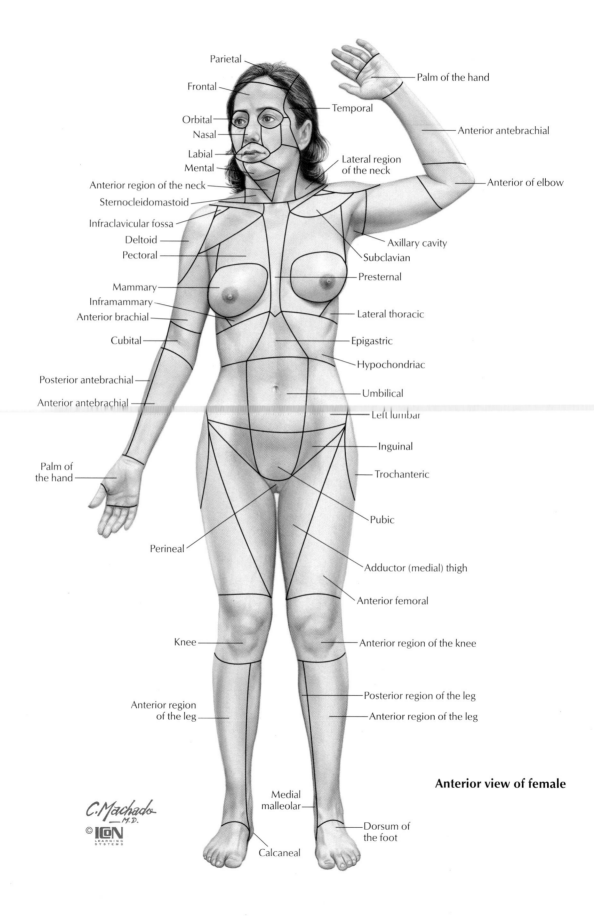

Parietal

Frontal

Orbital

Nasal

Labial

Mental

Anterior region of the neck

Sternocleidomastoid

Infraclavicular fossa

Deltoid

Pectoral

Mammary

Inframammary

Anterior brachial

Cubital

Posterior antebrachial

Anterior antebrachial

Palm of
the hand

Perineal

Palm of the hand

Temporal

Anterior antebrachial

Anterior of elbow

Lateral region
of the neck

Axillary cavity

Subclavian

Presternal

Lateral thoracic

Epigastric

Hypochondriac

Umbilical

Left lumbar

Inguinal

Trochanteric

Pubic

Adductor (medial) thigh

Anterior femoral

Knee

Anterior region of the knee

Posterior region of the leg

Anterior region of the leg

Anterior region
of the leg

Anterior view of female

Medial
malleolar

Calcaneal

Dorsum of
the foot

C. Machado
M.D.

© ICON
LEARNING
SYSTEMS

4

Surface Anatomy

When viewing a person in anatomical position from the posterior (or back) side, you can visualize the descriptive anatomical regions of the limbs, back, and gluteal area. For the back, most of the descriptive regions are related to bony features (e.g., vertebral [spinal] region and scapular region).

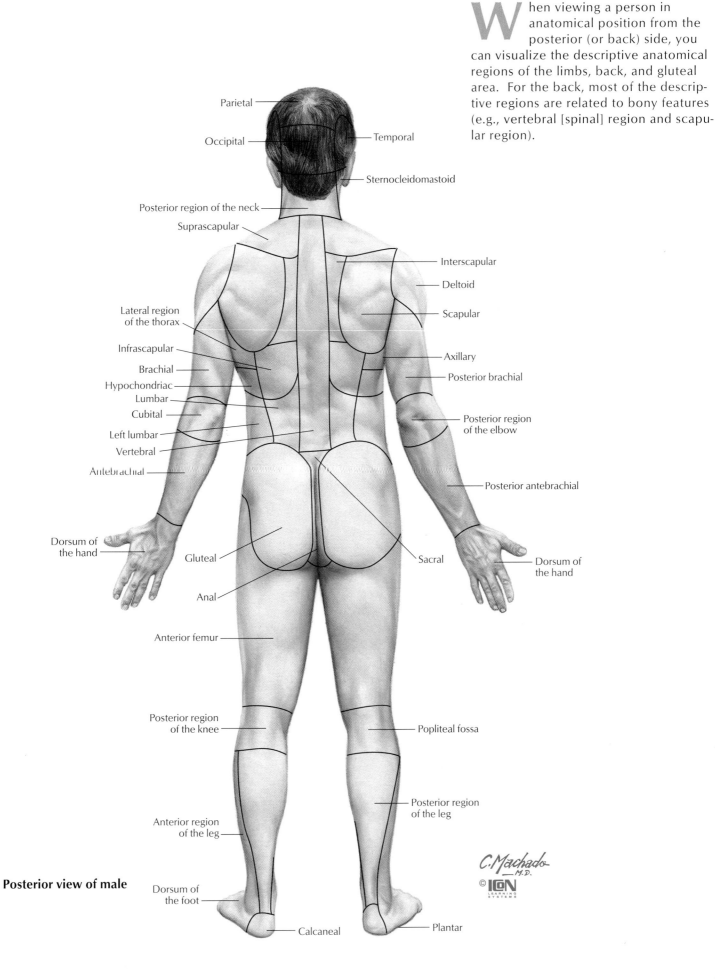

Parietal

Occipital — Temporal

Sternocleidomastoid

Posterior region of the neck

Suprascapular

Interscapular

Deltoid

Lateral region of the thorax

Scapular

Infrascapular

Axillary

Brachial

Posterior brachial

Hypochondriac

Lumbar

Cubital

Posterior region of the elbow

Left lumbar

Vertebral

Antebrachial

Posterior antebrachial

Dorsum of the hand

Gluteal

Sacral

Dorsum of the hand

Anal

Anterior femur

Posterior region of the knee

Popliteal fossa

Anterior region of the leg

Posterior region of the leg

Dorsum of the foot

Posterior view of male

Calcaneal

Plantar

5

Skin & Nails

Your skin is your body's largest organ, weighing about 9 lb on average. It provides protection from potentially lethal bacteria and viruses and shields you from the sun's ultraviolet rays. The skin is also a manufacturing plant, using the sun's energy to make vitamin D, which is essential in making bones strong.

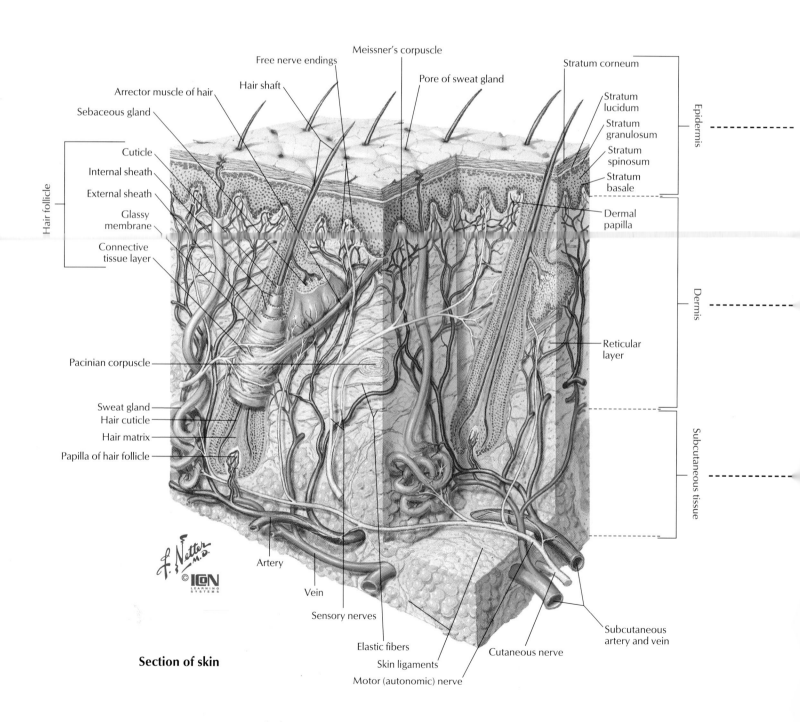

Section of skin

6

Nails & Hair

The outermost layer of skin is called the **epidermis** and is about as thick as a piece of paper. The top portion of the epidermis is known as the **stratum corneum**. It is composed of cells called **keratinocytes** that produce a tough protein called **keratin**, forming a flexible outer shield. The keratinocytes die as younger living cells from the lower part of the epidermis rise to the surface. Finally, they are rubbed off or fall off as new cells rise from below. This continual cycle renews your skin about once a month. Pigmented cells called **melanocytes** are located at the bottom layer of the epidermis. These cells produce the **melanin**, or pigment, that colors skin and helps protect against ultraviolet radiation. When exposed to sunlight, the melanocytes churn out more melanin, and your skin darkens to help shield against further damage.

Directly beneath the epidermis is the **dermis**, a thicker layer that contains collagen, blood and lymph vessels, nerves, hair follicles, and glands that produce sweat and oil. Blood vessels in the dermis expand or contract to maintain a constant body temperature. Cells called **fibroblasts** secrete **collagen**, which gives your skin its strength and firmness. **Elastin fibers** made of protein are contained in the dermis to give skin its elasticity.

The **subcutaneous tissue**, which consists of connective tissue and fat, is situated between the dermis and the underlying muscles or bones. It too contains blood vessels and infection-fighting white blood cells, but not to the same extent as in the dermis. Fat in the subcutaneous layer stores nutrients and insulates and cushions muscles and bones.

Your nails are a thickened, hardened form of epidermis. Nail cells originate from the base of the nail bed. They die quickly, but, unlike the keratinocytes, they are not sloughed off. They are also made of a much stronger form of keratin. Therefore, your nail is simply a much harder and thicker sheet of keratin than the topmost layer of skin. Hair, however, is a thin fiber made of many overlapping layers of keratin, which is produced in the hair root. (**Also see page 6.**)

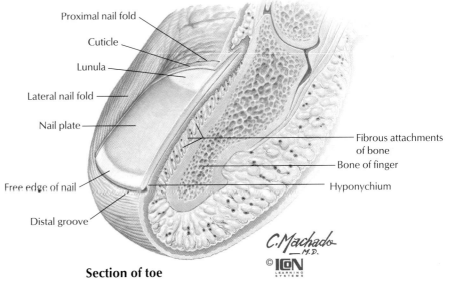

Proximal nail fold
Cuticle
Lunula
Lateral nail fold
Nail plate
Free edge of nail
Distal groove

Fibrous attachments of bone
Bone of finger
Hyponychium

C. Machado — M.D.

Section of toe

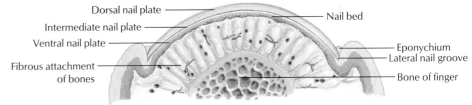

Dorsal nail plate
Intermediate nail plate
Ventral nail plate
Fibrous attachment of bones

Nail bed
Eponychium
Lateral nail groove
Bone of finger

Cross section of nail

Nail growth

The average growth rate of toenails is about 1 mm a month.

The rounded shape of the free edge of the nails is dictated by the shape of the lunula. After avulsion of a nail, the free edge of the new one grows parallel to the lunula.

The
Skeletal System

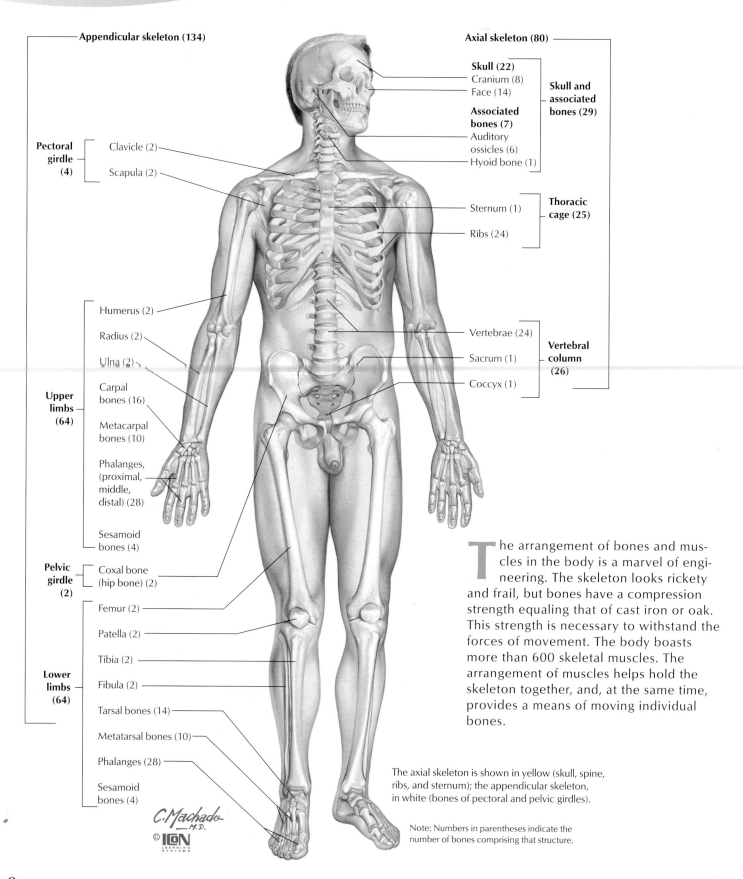

Appendicular skeleton (134)

Pectoral girdle (4)
- Clavicle (2)
- Scapula (2)

Upper limbs (64)
- Humerus (2)
- Radius (2)
- Ulna (2)
- Carpal bones (16)
- Metacarpal bones (10)
- Phalanges, (proximal, middle, distal) (28)
- Sesamoid bones (4)

Pelvic girdle (2)
- Coxal bone (hip bone) (2)

Lower limbs (64)
- Femur (2)
- Patella (2)
- Tibia (2)
- Fibula (2)
- Tarsal bones (14)
- Metatarsal bones (10)
- Phalanges (28)
- Sesamoid bones (4)

Axial skeleton (80)

Skull (22)
- Cranium (8)
- Face (14)

Associated bones (7)
- Auditory ossicles (6)
- Hyoid bone (1)

Skull and associated bones (29)

Thoracic cage (25)
- Sternum (1)
- Ribs (24)

- Vertebrae (24)
- Sacrum (1)
- Coccyx (1)

Vertebral column (26)

The arrangement of bones and muscles in the body is a marvel of engineering. The skeleton looks rickety and frail, but bones have a compression strength equaling that of cast iron or oak. This strength is necessary to withstand the forces of movement. The body boasts more than 600 skeletal muscles. The arrangement of muscles helps hold the skeleton together, and, at the same time, provides a means of moving individual bones.

The axial skeleton is shown in yellow (skull, spine, ribs, and sternum); the appendicular skeleton, in white (bones of pectoral and pelvic girdles).

Note: Numbers in parentheses indicate the number of bones comprising that structure.

C. Machado M.D.

© ICON
LEARNING
SYSTEMS

Structure of Bone

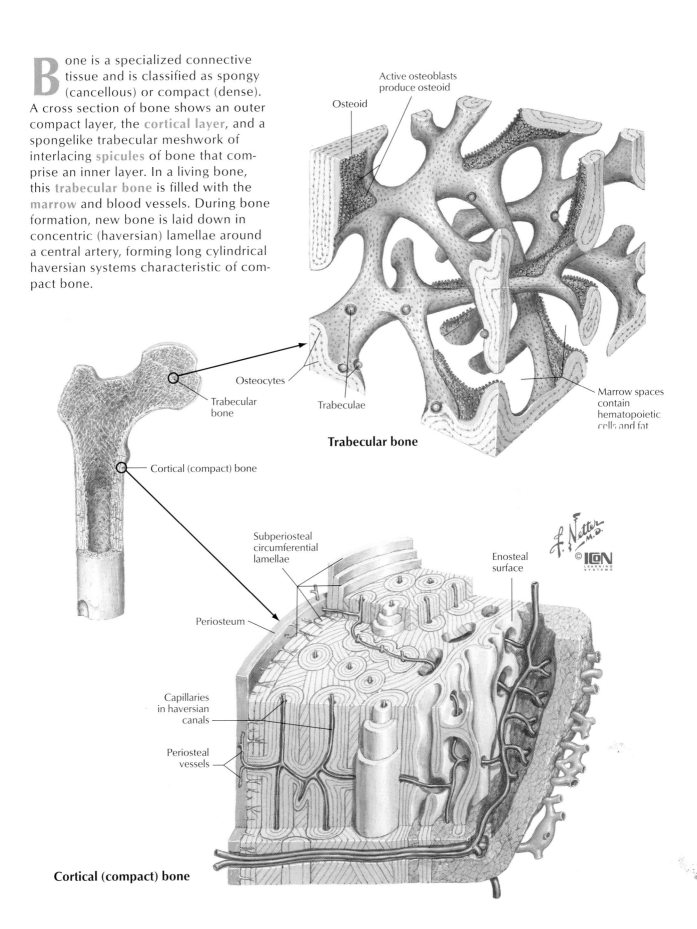

Bone is a specialized connective tissue and is classified as spongy (cancellous) or compact (dense). A cross section of bone shows an outer compact layer, the cortical layer, and a spongelike trabecular meshwork of interlacing spicules of bone that comprise an inner layer. In a living bone, this trabecular bone is filled with the marrow and blood vessels. During bone formation, new bone is laid down in concentric (haversian) lamellae around a central artery, forming long cylindrical haversian systems characteristic of compact bone.

Active osteoblasts produce osteoid

Osteoid

Osteocytes

Trabecular bone

Trabeculae

Marrow spaces contain hematopoietic cells and fat

Trabecular bone

Cortical (compact) bone

Subperiosteal circumferential lamellae

Enosteal surface

Periosteum

Capillaries in haversian canals

Periosteal vessels

Cortical (compact) bone

The Skeletal System
Bones of the Skull

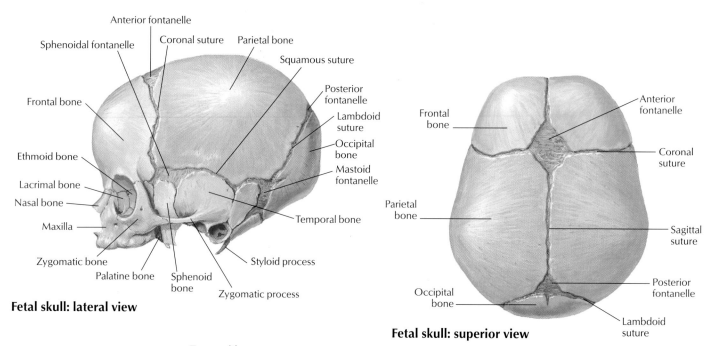

Fetal skull: lateral view

Anterior fontanelle
Sphenoidal fontanelle
Coronal suture
Parietal bone
Squamous suture
Posterior fontanelle
Lambdoid suture
Occipital bone
Mastoid fontanelle
Temporal bone
Styloid process
Zygomatic process
Sphenoid bone
Palatine bone
Zygomatic bone
Maxilla
Nasal bone
Lacrimal bone
Ethmoid bone
Frontal bone

Fetal skull: superior view

Frontal bone
Anterior fontanelle
Coronal suture
Sagittal suture
Posterior fontanelle
Lambdoid suture
Occipital bone
Parietal bone

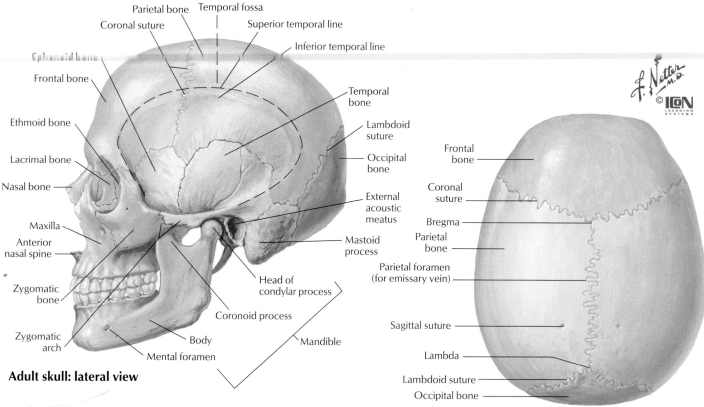

Adult skull: lateral view

Parietal bone
Temporal fossa
Coronal suture
Superior temporal line
Inferior temporal line
Temporal bone
Lambdoid suture
Occipital bone
External acoustic meatus
Mastoid process
Head of condylar process
Mandible
Coronoid process
Body
Mental foramen
Zygomatic arch
Zygomatic bone
Anterior nasal spine
Maxilla
Nasal bone
Lacrimal bone
Ethmoid bone
Frontal bone
Sphenoid bone

Adult skull: superior view

Frontal bone
Coronal suture
Bregma
Parietal bone
Parietal foramen (for emissary vein)
Sagittal suture
Lambda
Lambdoid suture
Occipital bone

The skull and its associated bones number 29 distinct bones and include the bones forming the **cranium** (flat bones joined by suture joints), the bones of the face, the middle ear ossicles and the lower jaw (**mandible**). During the fetal period and early infancy, the bones are still growing and fusing and are joined by membranous **fontanelles**, which demarcate the future site of cranial fusion (**sutures**).

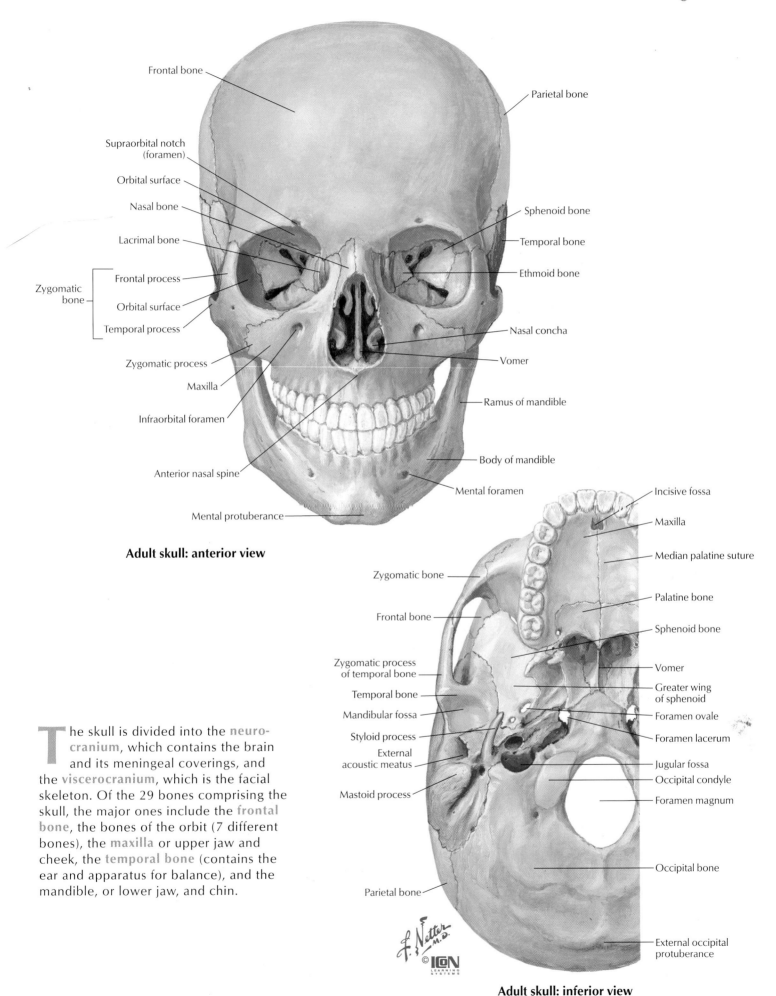

Frontal bone

Parietal bone

Supraorbital notch
(foramen)

Orbital surface

Nasal bone

Sphenoid bone

Lacrimal bone

Temporal bone

Zygomatic
bone

Frontal process

Ethmoid bone

Orbital surface

Temporal process

Zygomatic process

Nasal concha

Maxilla

Vomer

Infraorbital foramen

Ramus of mandible

Anterior nasal spine

Body of mandible

Mental foramen

Mental protuberance

Adult skull: anterior view

Incisive fossa

Maxilla

Median palatine suture

Zygomatic bone

Palatine bone

Frontal bone

Sphenoid bone

Vomer

Zygomatic process
of temporal bone

Greater wing
of sphenoid

Temporal bone

Foramen ovale

Mandibular fossa

Foramen lacerum

Styloid process

Jugular fossa

External
acoustic meatus

Occipital condyle

Foramen magnum

Mastoid process

Occipital bone

Parietal bone

The skull is divided into the **neuro-cranium**, which contains the brain and its meningeal coverings, and the **viscerocranium**, which is the facial skeleton. Of the 29 bones comprising the skull, the major ones include the **frontal bone**, the bones of the orbit (7 different bones), the **maxilla** or upper jaw and cheek, the **temporal bone** (contains the ear and apparatus for balance), and the mandible, or lower jaw, and chin.

External occipital
protuberance

Adult skull: inferior view

11

Sagittal Section of Skull

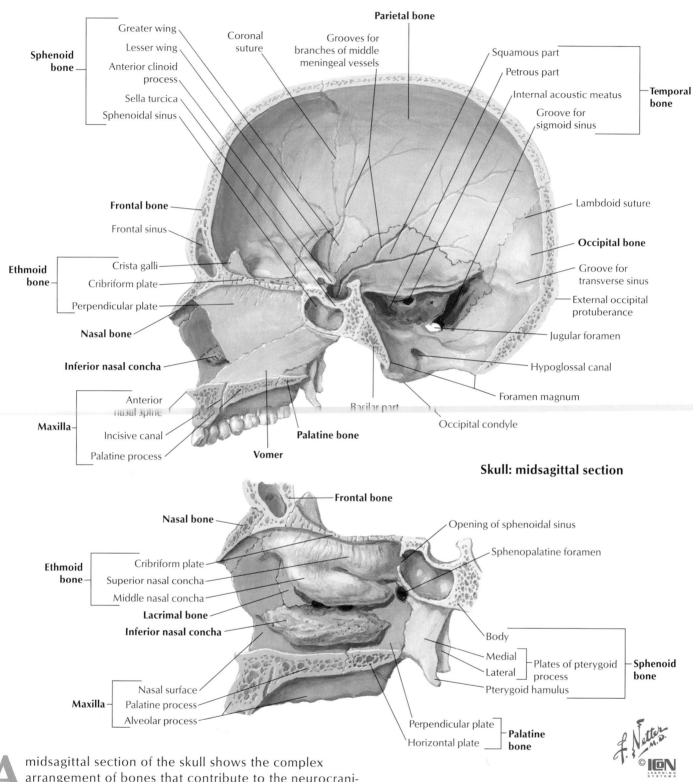

Sphenoid bone
- Greater wing
- Lesser wing
- Anterior clinoid process
- Sella turcica
- Sphenoidal sinus

Coronal suture

Grooves for branches of middle meningeal vessels

Parietal bone

Squamous part
Petrous part
Internal acoustic meatus
Groove for sigmoid sinus
— Temporal bone

Frontal bone
Frontal sinus

Lambdoid suture

Ethmoid bone
- Crista galli
- Cribriform plate
- Perpendicular plate

Occipital bone

Groove for transverse sinus

External occipital protuberance

Nasal bone

Jugular foramen

Inferior nasal concha

Hypoglossal canal

Maxilla
- Anterior nasal spine
- Incisive canal
- Palatine process

Basilar part

Palatine bone

Vomer

Foramen magnum

Occipital condyle

Skull: midsagittal section

Frontal bone

Nasal bone

Opening of sphenoidal sinus

Sphenopalatine foramen

Ethmoid bone
- Cribriform plate
- Superior nasal concha
- Middle nasal concha
- Lacrimal bone
- Inferior nasal concha

Body

Medial
Lateral
— Plates of pterygoid process
Pterygoid hamulus
— Sphenoid bone

Maxilla
- Nasal surface
- Palatine process
- Alveolar process

Perpendicular plate
Horizontal plate
— Palatine bone

Nasal cavity: midsagittal section

A midsagittal section of the skull shows the complex arrangement of bones that contribute to the neurocranium (holds the brain) and the nasal cavity, hard palate, and upper jaw. The view of the nasal cavity shows the lateral nasal wall and the concha (nasal turbinates) that are normally covered with a mucosa rich in blood vessels and fine hairs, which warm, humidify, and filter the air you breathe.

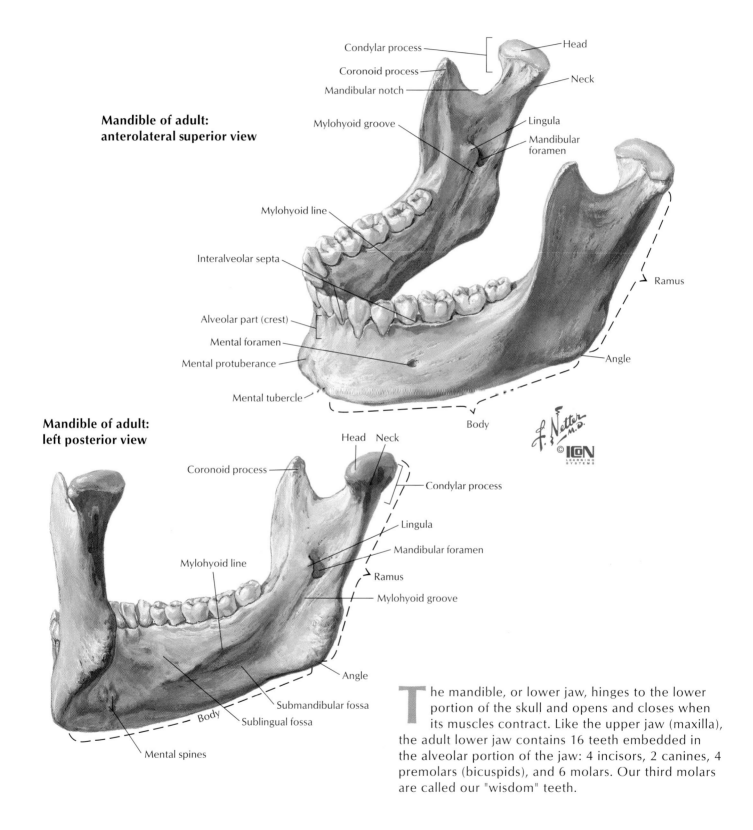

**Mandible of adult:
anterolateral superior view**

Condylar process — Head
Coronoid process — Neck
Mandibular notch
Mylohyoid groove — Lingula
— Mandibular foramen
Mylohyoid line
Interalveolar septa
— Ramus
Alveolar part (crest)
Mental foramen
Mental protuberance — Angle
Mental tubercle
Body

**Mandible of adult:
left posterior view**

Head Neck
Coronoid process — Condylar process
— Lingula
— Mandibular foramen
Mylohyoid line — Ramus
— Mylohyoid groove
Angle
Submandibular fossa
Body Sublingual fossa
Mental spines

The mandible, or lower jaw, hinges to the lower portion of the skull and opens and closes when its muscles contract. Like the upper jaw (maxilla), the adult lower jaw contains 16 teeth embedded in the alveolar portion of the jaw: 4 incisors, 2 canines, 4 premolars (bicuspids), and 6 molars. Our third molars are called our "wisdom" teeth.

The Skeletal System
Vertebrae – Neck

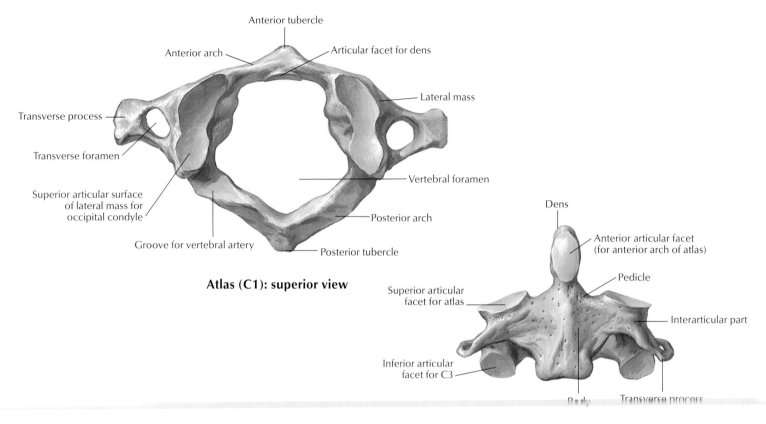

Atlas (C1): superior view

- Anterior tubercle
- Anterior arch
- Articular facet for dens
- Lateral mass
- Transverse process
- Transverse foramen
- Superior articular surface of lateral mass for occipital condyle
- Vertebral foramen
- Groove for vertebral artery
- Posterior arch
- Posterior tubercle

Axis (C2): anterior view

- Dens
- Anterior articular facet (for anterior arch of atlas)
- Superior articular facet for atlas
- Pedicle
- Interarticular part
- Inferior articular facet for C3
- Body
- Transverse process

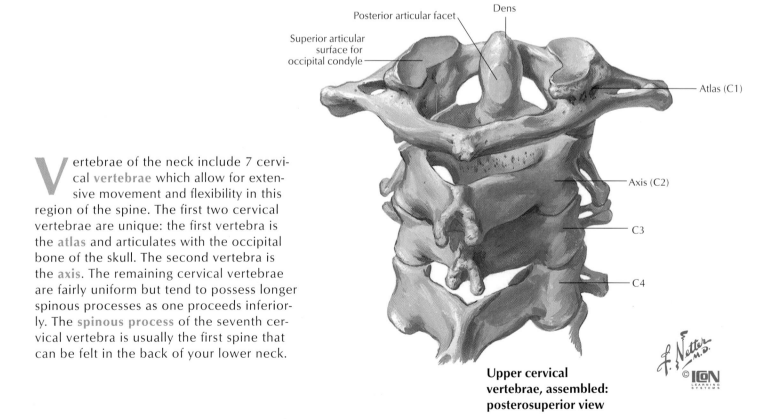

Upper cervical vertebrae, assembled: posterosuperior view

- Posterior articular facet
- Dens
- Superior articular surface for occipital condyle
- Atlas (C1)
- Axis (C2)
- C3
- C4

Vertebrae of the neck include 7 cervical **vertebrae** which allow for extensive movement and flexibility in this region of the spine. The first two cervical vertebrae are unique: the first vertebra is the **atlas** and articulates with the occipital bone of the skull. The second vertebra is the **axis**. The remaining cervical vertebrae are fairly uniform but tend to possess longer spinous processes as one proceeds inferiorly. The **spinous process** of the seventh cervical vertebra is usually the first spine that can be felt in the back of your lower neck.

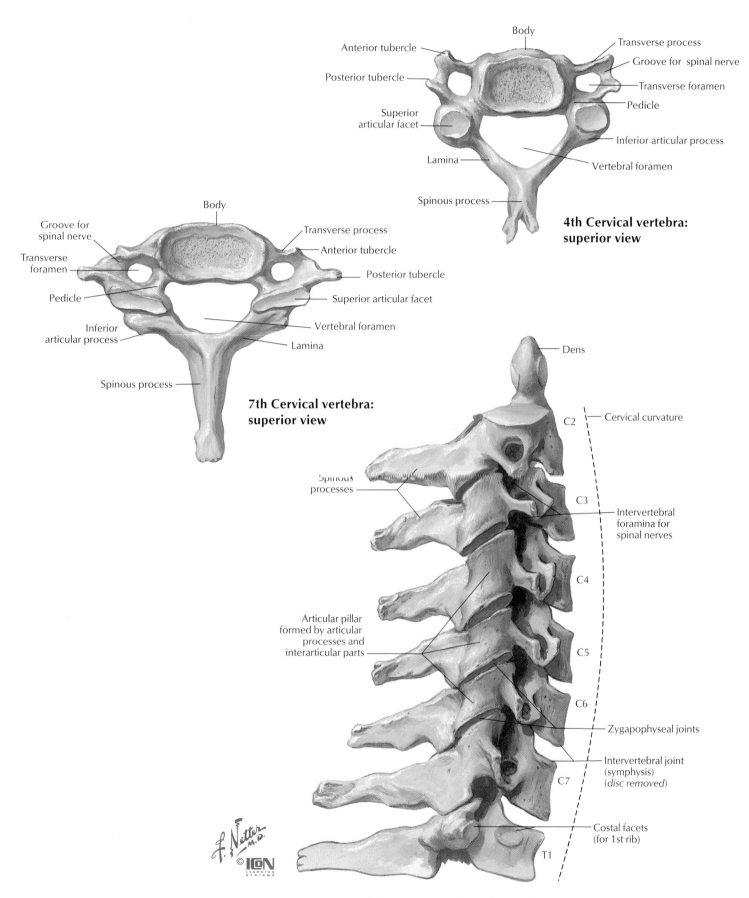

Body

Anterior tubercle

Posterior tubercle

Superior
articular facet

Lamina

Spinous process

Transverse process

Groove for spinal nerve

Transverse foramen

Pedicle

Inferior articular process

Vertebral foramen

**4th Cervical vertebra:
superior view**

Body

Groove for
spinal nerve

Transverse
foramen

Pedicle

Inferior
articular process

Spinous process

Transverse process

Anterior tubercle

Posterior tubercle

Superior articular facet

Vertebral foramen

Lamina

**7th Cervical vertebra:
superior view**

Dens

Spinous
processes

Articular pillar
formed by articular
processes and
interarticular parts

C2

C3

C4

C5

C6

C7

T1

Cervical curvature

Intervertebral
foramina for
spinal nerves

Zygapophyseal joints

Intervertebral joint
(symphysis)
(*disc removed*)

Costal facets
(for 1st rib)

**2nd Cervical to 1st thoracic vertebrae:
right lateral view**

F. Netter
M.D.

© ICON
LEARNING
SYSTEMS

Vertebrae – Back

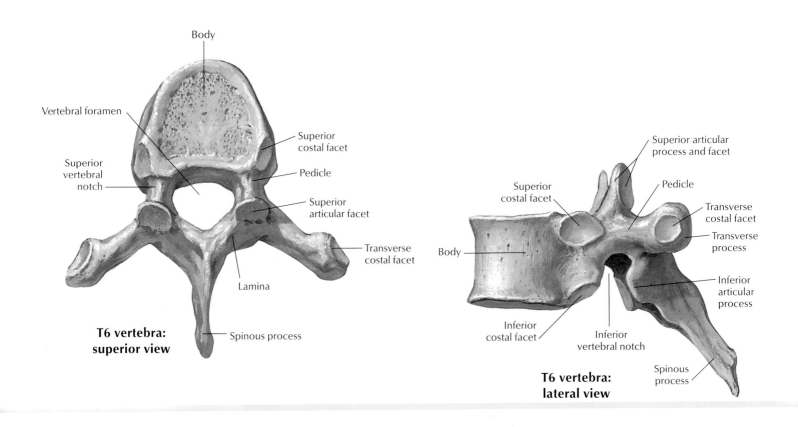

Body

Vertebral foramen

Superior
vertebral
notch

Superior
costal facet

Pedicle

Superior
articular facet

Transverse
costal facet

Lamina

Spinous process

**T6 vertebra:
superior view**

Superior
costal facet

Body

Inferior
costal facet

Superior articular
process and facet

Pedicle

Transverse
costal facet

Transverse
process

Inferior
articular
process

Inferior
vertebral notch

Spinous
process

**T6 vertebra:
lateral view**

The remaining vertebrae of the spine below the cervical level include 12 **thoracic vertebrae** (with long spinous processes and facets for rib articulation), 5 **lumbar vertebrae** for strong support of the trunk, 5 fused **sacral vertebrae** for articulation with the pelvic girdle, and 4 fused **coccygeal vertebrae** (represent the remnant of our tailbone).

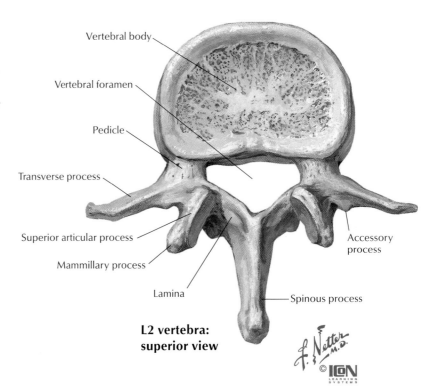

Vertebral body

Vertebral foramen

Pedicle

Transverse process

Superior articular process

Mammillary process

Lamina

Spinous process

Accessory
process

**L2 vertebra:
superior view**

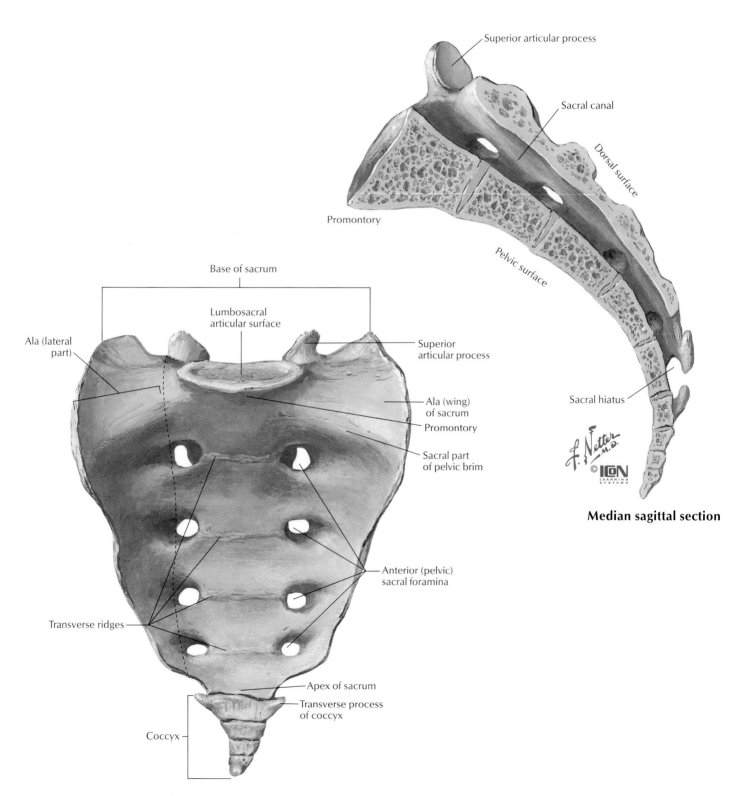

Superior articular process

Sacral canal

Dorsal surface

Promontory

Pelvic surface

Sacral hiatus

Median sagittal section

Base of sacrum

Lumbosacral articular surface

Ala (lateral part)

Superior articular process

Ala (wing) of sacrum

Promontory

Sacral part of pelvic brim

Anterior (pelvic) sacral foramina

Transverse ridges

Apex of sacrum

Transverse process of coccyx

Coccyx

Pelvic surface: anterior inferior view

The Skeletal System
Spine

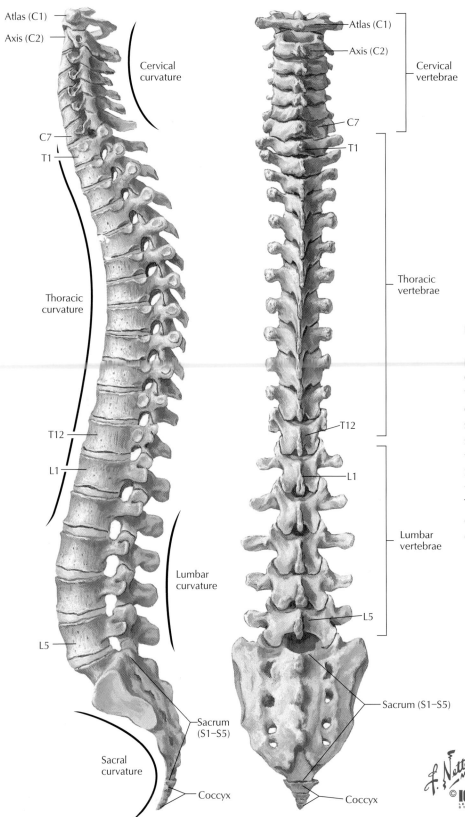

Atlas (C1)
Axis (C2)

Cervical curvature

C7
T1

Thoracic curvature

T12

L1

Lumbar curvature

L5

Sacral curvature

Sacrum (S1–S5)

Coccyx

Atlas (C1)
Axis (C2)

Cervical vertebrae

C7
T1

Thoracic vertebrae

T12

L1

Lumbar vertebrae

L5

Sacrum (S1–S5)

Coccyx

nterlocking bones called **vertebrae** form the spine. Each vertebra has a roughly cylindrical body with a bony ring attached to its back surface. **Processes**, or posterior elements, that project in several directions are found on this ring. Vertebrae are stacked on top of each other and are linked together by ligaments and joints that permit a small amount of forward, backward, and side-to-side bending, as well as some twisting and up-and-down movement. **Spinal ligaments** and muscles attach to the **spinous** and **transverse processes**. The ligaments connect the vertebrae to one another, making the spine a flexible column.

Spine: lateral and posterior views

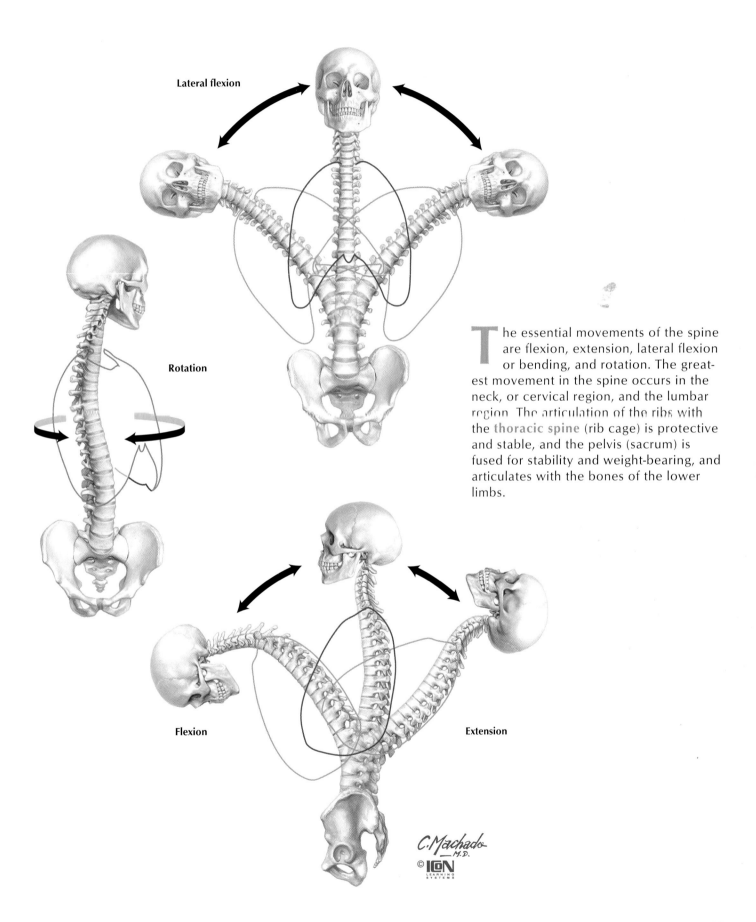

Lateral flexion

Rotation

T he essential movements of the spine are flexion, extension, lateral flexion or bending, and rotation. The greatest movement in the spine occurs in the neck, or cervical region, and the lumbar region. The articulation of the ribs with the **thoracic spine** (rib cage) is protective and stable, and the pelvis (sacrum) is fused for stability and weight-bearing, and articulates with the bones of the lower limbs.

Flexion

Extension

C. Machado
M.D.
© ICON
LEARNING
SYSTEMS

19

The Skeletal System
Thorax

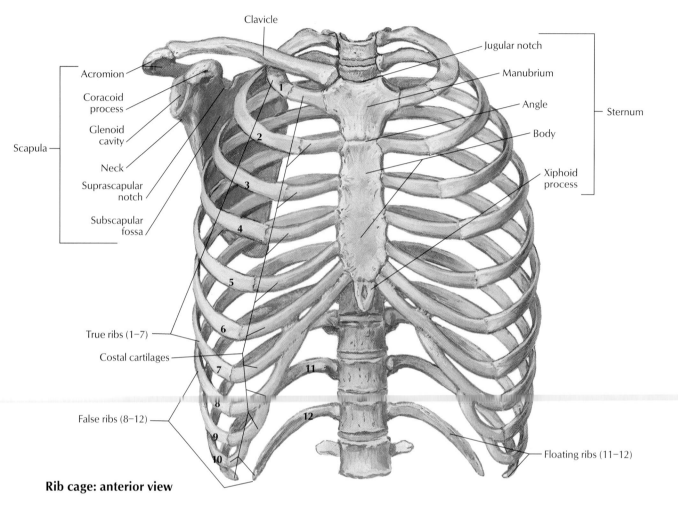

Clavicle

Acromion
Coracoid process
Glenoid cavity
Scapula
Neck
Suprascapular notch
Subscapular fossa

Jugular notch
Manubrium
Angle
Body
Sternum
Xiphoid process

1
2
3
4
5
6
7
8
9
10
11
12

True ribs (1–7)
Costal cartilages
False ribs (8–12)

Floating ribs (11–12)

Rib cage: anterior view

The **thoracic cage** is part of the axial skeleton and includes the thoracic vertebrae, **sternum**, and **ribs**. This arrangement of bones protects the vital structures of the **thoracic cavity**, which includes the lungs, the heart, and the great vessels entering or leaving the heart. Humans have 12 pairs of ribs: ribs 1–7 articulate with the sternum directly and are called *true* ribs; ribs 8–10 articulate with costal cartilages of the rib above and are called *false* ribs, and ribs 11–12 articulate with thoracic vertebrae only and are called *floating* ribs (and also may be included in the designation as false ribs, see image).

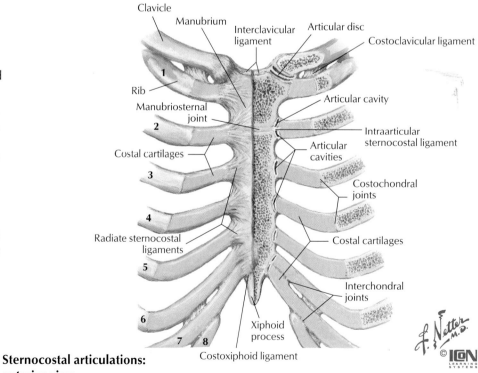

Clavicle
Manubrium
Interclavicular ligament
Articular disc
Costoclavicular ligament

1
Rib
Manubriosternal joint
2
Costal cartilages
3
4
Radiate sternocostal ligaments
5
6
7   8
Xiphoid process
Costoxiphoid ligament

Articular cavity
Intraarticular sternocostal ligament
Articular cavities
Costochondral joints
Costal cartilages
Interchondral joints

**Sternocostal articulations:
anterior view**

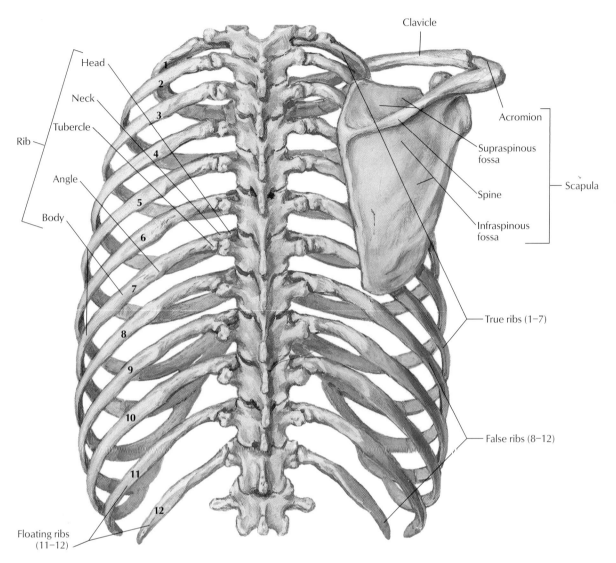

Clavicle

Head

Neck

Tubercle

Rib

Angle

Body

Acromion

Supraspinous
fossa

Spine

Scapula

Infraspinous
fossa

True ribs (1–7)

False ribs (8–12)

Floating ribs
(11–12)

Rib cage: posterior view

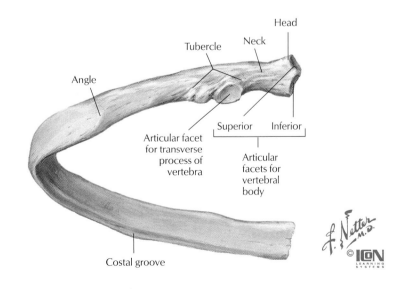

Head

Neck

Tubercle

Angle

Superior   Inferior

Articular facet
for transverse
process of
vertebra

Articular
facets for
vertebral
body

Costal groove

Middle rib: posterior view

The Skeletal System
Pelvis

T he **pelvis** is made up of four bones: two hip bones, the curved triangular bone at the base of the spine (**sacrum**), and the tailbone (**coccyx**). The hip joint is where the thighbones (**femurs**) attach to the pelvis.

The hip is designed to support the full weight of the body while allowing movement in all directions. To accomplish this, the top of the thighbone is shaped in a smooth ball that fits snugly within the **acetabulum**, a deep socket in the pelvis. The hip joint is a ball-and-socket joint reinforced by a strong ring of cartilage (**labrum**). A strong joint capsule of four ligaments allows the hip a wide range of motion while bearing the full weight of the upper body.

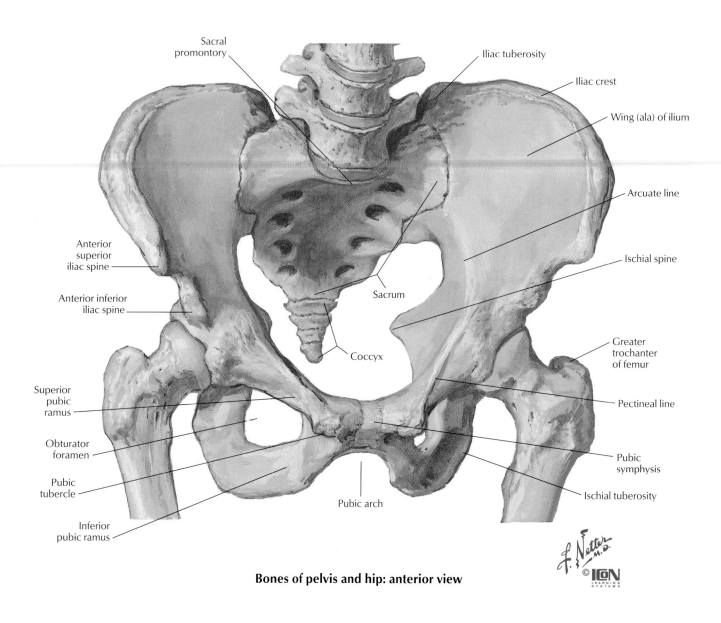

Sacral promontory

Iliac tuberosity

Iliac crest

Wing (ala) of ilium

Arcuate line

Anterior superior iliac spine

Ischial spine

Anterior inferior iliac spine

Sacrum

Greater trochanter of femur

Coccyx

Superior pubic ramus

Pectineal line

Obturator foramen

Pubic symphysis

Pubic tubercle

Ischial tuberosity

Pubic arch

Inferior pubic ramus

Bones of pelvis and hip: anterior view

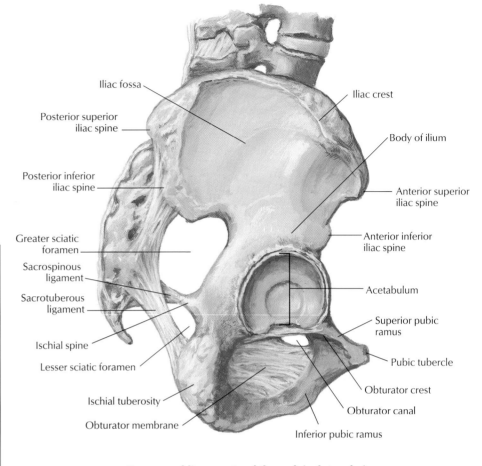

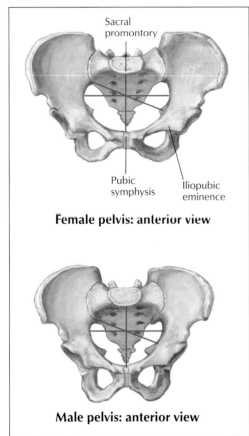

Sacral
promontory

Pubic
symphysis

Iliopubic
eminence

Female pelvis: anterior view

Male pelvis: anterior view

Iliac fossa

Posterior superior
iliac spine

Posterior inferior
iliac spine

Greater sciatic
foramen

Sacrospinous
ligament

Sacrotuberous
ligament

Ischial spine

Lesser sciatic foramen

Ischial tuberosity

Obturator membrane

Iliac crest

Body of ilium

Anterior superior
iliac spine

Anterior inferior
iliac spine

Acetabulum

Superior pubic
ramus

Pubic tubercle

Obturator crest

Obturator canal

Inferior pubic ramus

Bones and ligaments of the pelvis: lateral view

The differences evident in the male
and the female pelves are related to
their functional anatomy. Compared
with the male pelvis, the female pelvis generally has smaller bones, an oval rather
than a heart-shaped inlet, a larger pelvic
outlet for childbirth, a wider pelvic cavity
with flared iliac fossae, and a shorter,
wider sacrum.

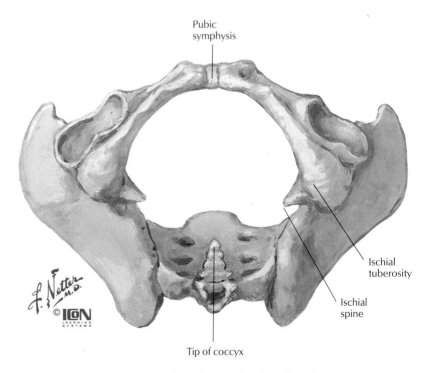

Pubic
symphysis

Ischial
tuberosity

Ischial
spine

Tip of coccyx

Female pelvic outlet: inferior view

The Skeletal System
Upper Limb

The bones of the shoulder include the collarbone (clavicle), the shoulder blade (scapula), and the bone of the upper arm (humerus). The spine and the rib cage form the upper body.

The lower arm has two bones, the radius and the ulna. The upper and lower arms meet at the elbow joint.

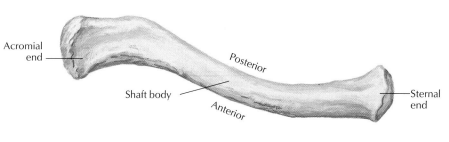

Acromial end

Posterior

Shaft body

Anterior

Sternal end

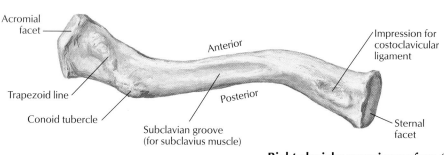

Acromial facet

Anterior

Impression for costoclavicular ligament

Trapezoid line

Conoid tubercle

Posterior

Subclavian groove (for subclavius muscle)

Sternal facet

Right clavicle: superior surface (top), inferior surface (bottom)

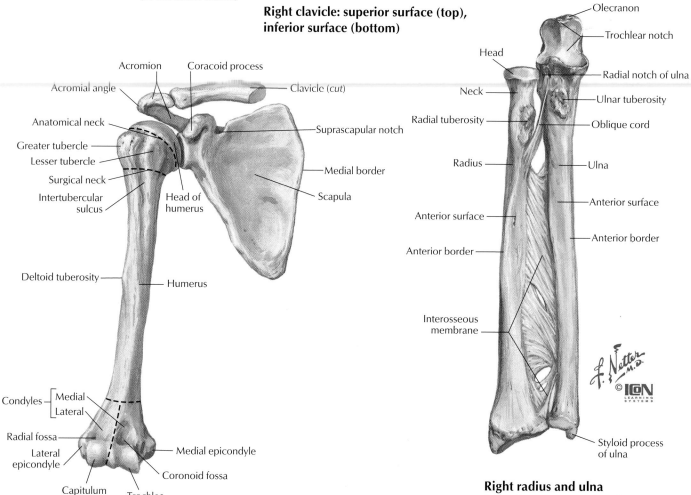

Acromion

Coracoid process

Acromial angle

Clavicle (cut)

Anatomical neck

Greater tubercle

Lesser tubercle

Surgical neck

Suprascapular notch

Intertubercular sulcus

Head of humerus

Medial border

Scapula

Deltoid tuberosity

Humerus

Condyles — Medial / Lateral

Radial fossa

Lateral epicondyle

Medial epicondyle

Capitulum

Coronoid fossa

Trochlea

Right arm (humerus) and shoulder joint (humerus, clavicle, and scapula): anterior view

Olecranon

Trochlear notch

Head

Radial notch of ulna

Neck

Ulnar tuberosity

Radial tuberosity

Oblique cord

Radius

Ulna

Anterior surface

Anterior surface

Anterior border

Anterior border

Interosseous membrane

Styloid process of ulna

Right radius and ulna in supination: anterior view

24

Hand

The wrist is made up of eight small bones called carpal bones. The bones of the fingers include the metacarpals and phalanges (the bones that extend from the knuckles to the fingertips).

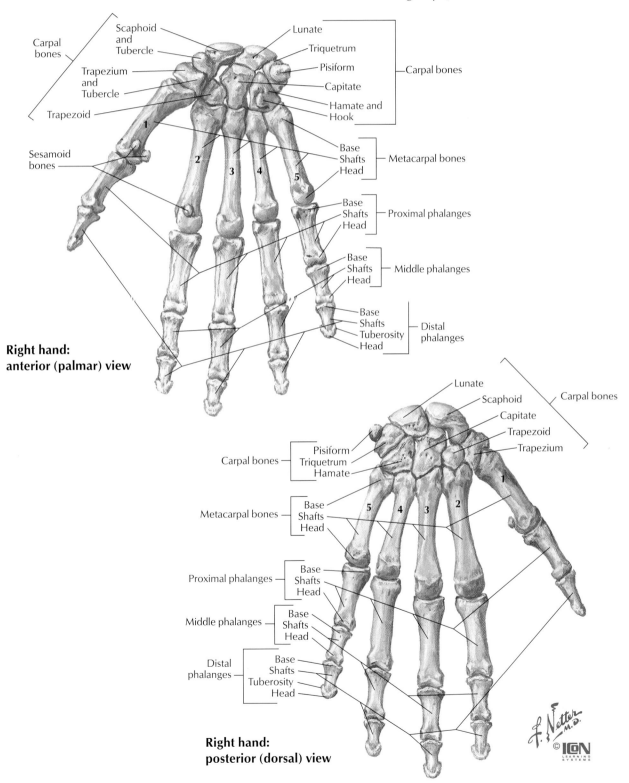

Carpal bones

Scaphoid and Tubercle

Lunate

Triquetrum

Trapezium and Tubercle

Pisiform

Capitate

Trapezoid

Hamate and Hook

Carpal bones

Sesamoid bones

1

2

3

4

5

Base
Shafts
Head — Metacarpal bones

Base
Shafts
Head — Proximal phalanges

Base
Shafts
Head — Middle phalanges

Base
Shafts
Tuberosity
Head — Distal phalanges

**Right hand:
anterior (palmar) view**

Lunate

Scaphoid

Capitate

Trapezoid

Trapezium

Carpal bones

Pisiform
Triquetrum
Hamate

Carpal bones

1

2

3

4

5

Metacarpal bones

Base
Shafts
Head

Proximal phalanges

Base
Shafts
Head

Middle phalanges

Base
Shafts
Head

Distal phalanges

Base
Shafts
Tuberosity
Head

**Right hand:
posterior (dorsal) view**

25

The Skeletal System
Lower Limb

The bone of the thigh is called the **femur**, and the lower leg has two bones, the shin bone (**tibia**) and the calf (**fibula**).

The knee connects the thighbone and the shin bone, the larger front bone of the calf. The top of the tibia is rather flat with a middle bump.

The kneecap (**patella**) is a small, flat bone that floats in front of the knee joint. The patella moves within a groove between the two **condyles** (joint prominences of a bone, resembling knuckles) of the femur. It protects other knee structures and applies leverage to help straighten the joint. Although the knee moves in only one direction, it can also slightly rotate or move from side to side.

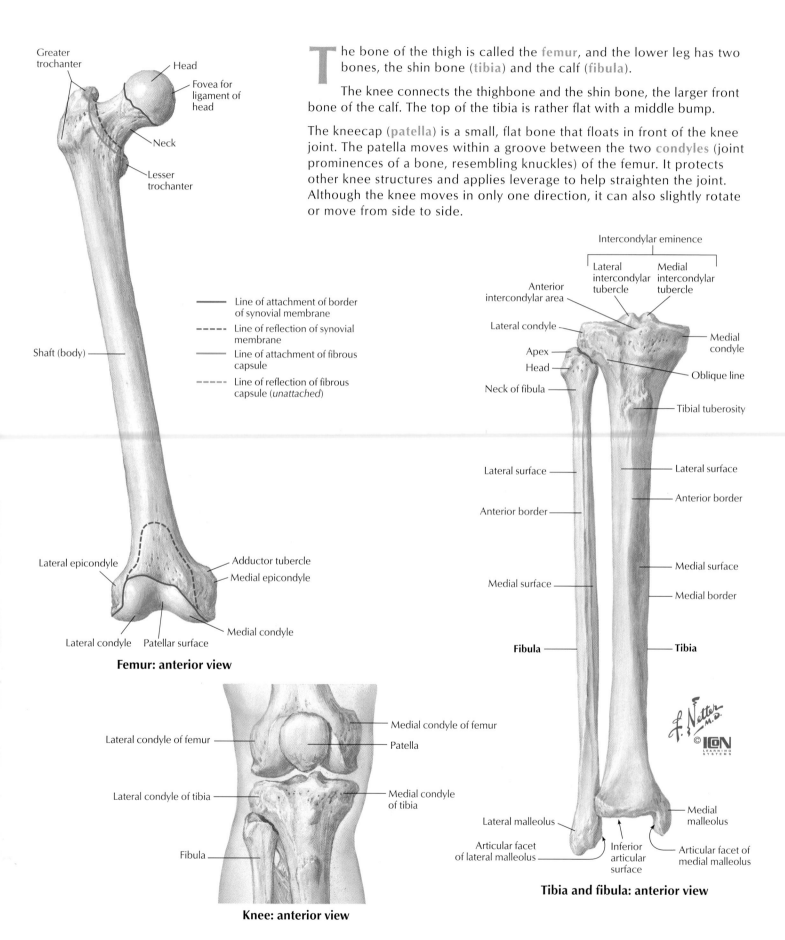

Greater trochanter
Head
Fovea for ligament of head
Neck
Lesser trochanter
Shaft (body)

——— Line of attachment of border of synovial membrane
---- Line of reflection of synovial membrane
——— Line of attachment of fibrous capsule
---- Line of reflection of fibrous capsule (*unattached*)

Lateral epicondyle
Adductor tubercle
Medial epicondyle
Lateral condyle
Patellar surface
Medial condyle

Femur: anterior view

Lateral condyle of femur
Medial condyle of femur
Patella
Lateral condyle of tibia
Medial condyle of tibia
Fibula

Knee: anterior view

Intercondylar eminence
Lateral intercondylar tubercle
Medial intercondylar tubercle
Anterior intercondylar area
Lateral condyle
Medial condyle
Apex
Head
Oblique line
Neck of fibula
Tibial tuberosity
Lateral surface
Lateral surface
Anterior border
Anterior border
Medial surface
Medial surface
Medial border
Fibula
Tibia
Lateral malleolus
Medial malleolus
Articular facet of lateral malleolus
Inferior articular surface
Articular facet of medial malleolus

Tibia and fibula: anterior view

Foot

T he shin bone and calf connect to the foot at the talus bone, a small bone located between the heel bone and the two bones of the lower leg and one of seven tarsal bones (bones of the heel and ankle). The tarsal bones, metatarsal (ball of the foot) bones, and phalanges (14 bones of the toes) are the bones of the foot and the toes. Hidden under the base of the big toe are two tiny bones called sesamoids.

The midfoot consists of five rectangular bones (three cuneiforms, the cuboid, and the navicular) that are short and broad and fit tightly together. The hindfoot contains the largest bones in the foot, including the heel bone, or calcaneus, and the talus.

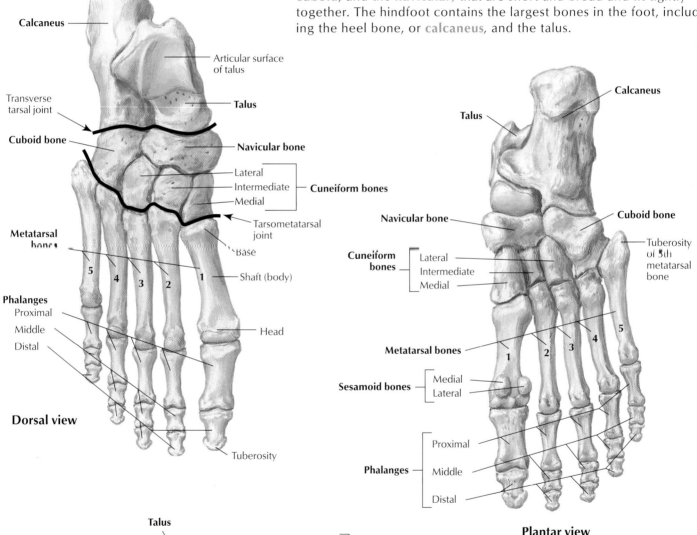

Calcaneus

Articular surface of talus

Transverse tarsal joint

Talus

Cuboid bone

Navicular bone

Lateral
Intermediate — Cuneiform bones
Medial

Tarsometatarsal joint

Metatarsal bones

Base

5 4 3 2 1 — Shaft (body)

Phalanges
Proximal
Middle
Distal

Head

Dorsal view

Calcaneus

Talus

Navicular bone

Cuboid bone

Cuneiform bones
Lateral
Intermediate
Medial

Tuberosity of 5th metatarsal bone

Metatarsal bones
1 2 3 4 5

Sesamoid bones
Medial
Lateral

Phalanges
Proximal
Middle
Distal

Plantar view

Tuberosity

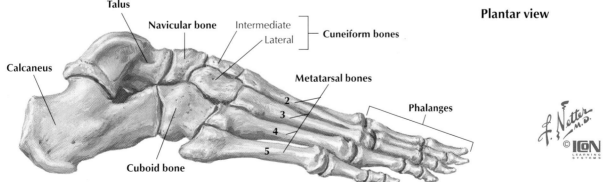

Talus

Navicular bone

Intermediate
Lateral — Cuneiform bones

Calcaneus

Metatarsal bones

2
3
4
5

Phalanges

Cuboid bone

Lateral view

The
Joints & Ligaments

Two or more bones articulate with one another at **joints**. Skeletal muscles act on these joints to bring about movement. Most of the joints in the human body are **synovial joints**, which means they have articular cartilage on their opposing surfaces and possess a synovial cavity that contains a small amount of synovial fluid to lubricate the joint, all surrounded by a capsule of tough connective tissue.

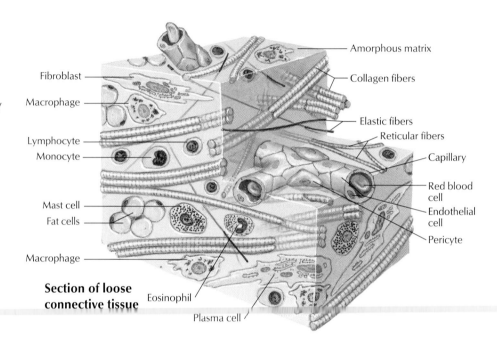

Fibroblast

Macrophage

Lymphocyte

Monocyte

Mast cell

Fat cells

Macrophage

Section of loose connective tissue

Eosinophil

Plasma cell

Amorphous matrix

Collagen fibers

Elastic fibers

Reticular fibers

Capillary

Red blood cell

Endothelial cell

Pericyte

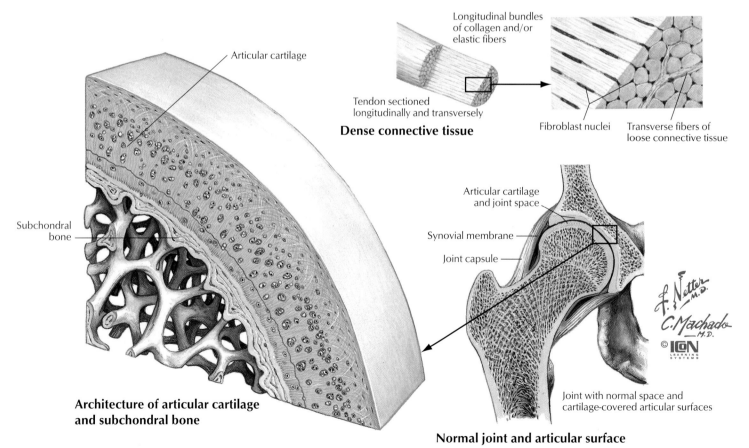

Articular cartilage

Subchondral bone

Architecture of articular cartilage and subchondral bone

Longitudinal bundles of collagen and/or elastic fibers

Tendon sectioned longitudinally and transversely

Dense connective tissue

Fibroblast nuclei

Transverse fibers of loose connective tissue

Articular cartilage and joint space

Synovial membrane

Joint capsule

Joint with normal space and cartilage-covered articular surfaces

Normal joint and articular surface

Types of Joints

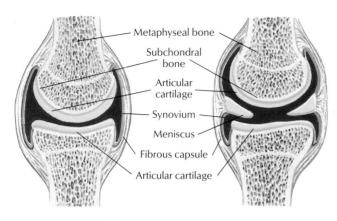

Structure of synovial joints

Synovial joints are the most common type of joint in humans and usually allow for considerable movement. Synovial joints have articular cartilage on their opposing surfaces and possess a synovial cavity, all surrounded by a fibrous capsule of connective tissue. Based on their shape and the movements they permit, synovial joints are classified into one of six types: hinge, pivot, saddle, condyloid, ball and socket, and plane.

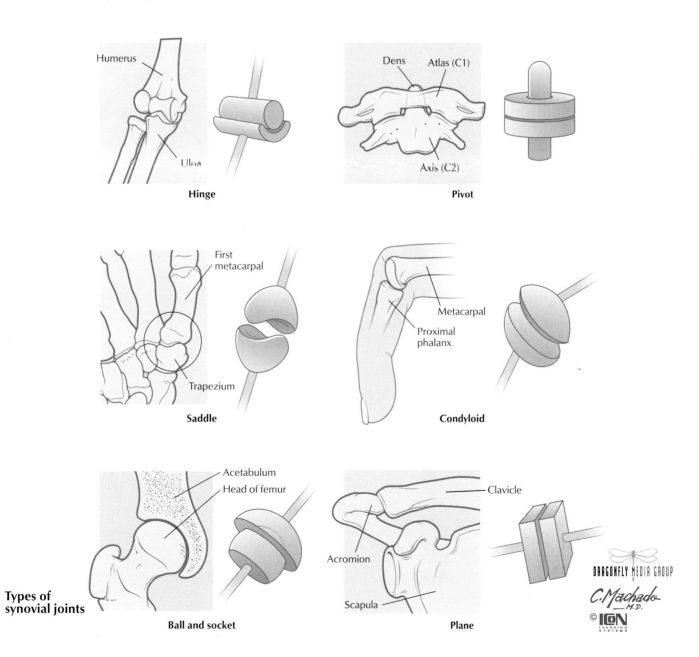

Hinge

Pivot

Saddle

Condyloid

Types of synovial joints

Ball and socket

Plane

DRAGONFLY MEDIA GROUP

C. Machado
—M.D.

© ICON
LEARNING SYSTEMS

The Joints & Ligaments

Shoulder

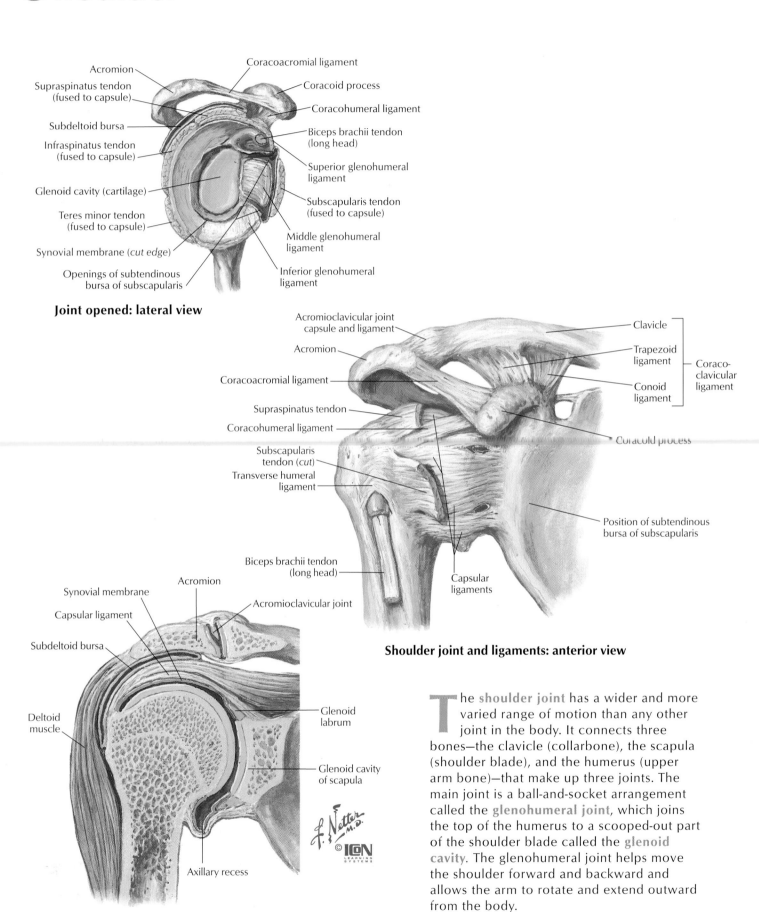

Acromion

Supraspinatus tendon (fused to capsule)

Subdeltoid bursa

Infraspinatus tendon (fused to capsule)

Glenoid cavity (cartilage)

Teres minor tendon (fused to capsule)

Synovial membrane (*cut edge*)

Openings of subtendinous bursa of subscapularis

Coracoacromial ligament

Coracoid process

Coracohumeral ligament

Biceps brachii tendon (long head)

Superior glenohumeral ligament

Subscapularis tendon (fused to capsule)

Middle glenohumeral ligament

Inferior glenohumeral ligament

Joint opened: lateral view

Acromioclavicular joint capsule and ligament

Acromion

Coracoacromial ligament

Supraspinatus tendon

Coracohumeral ligament

Subscapularis tendon (*cut*)

Transverse humeral ligament

Biceps brachii tendon (long head)

Capsular ligaments

Clavicle

Trapezoid ligament

Conoid ligament

Coraco-clavicular ligament

Coracoid process

Position of subtendinous bursa of subscapularis

Shoulder joint and ligaments: anterior view

Synovial membrane

Capsular ligament

Subdeltoid bursa

Deltoid muscle

Acromion

Acromioclavicular joint

Glenoid labrum

Glenoid cavity of scapula

Axillary recess

Coronal section through shoulder joint

The **shoulder joint** has a wider and more varied range of motion than any other joint in the body. It connects three bones—the clavicle (collarbone), the scapula (shoulder blade), and the humerus (upper arm bone)—that make up three joints. The main joint is a ball-and-socket arrangement called the **glenohumeral joint**, which joins the top of the humerus to a scooped-out part of the shoulder blade called the **glenoid cavity**. The glenohumeral joint helps move the shoulder forward and backward and allows the arm to rotate and extend outward from the body.

The elbow joint is a type of synovial hinge joint that provides for flexion and extension. It is composed of the humeroulnar and humeroradial joints.

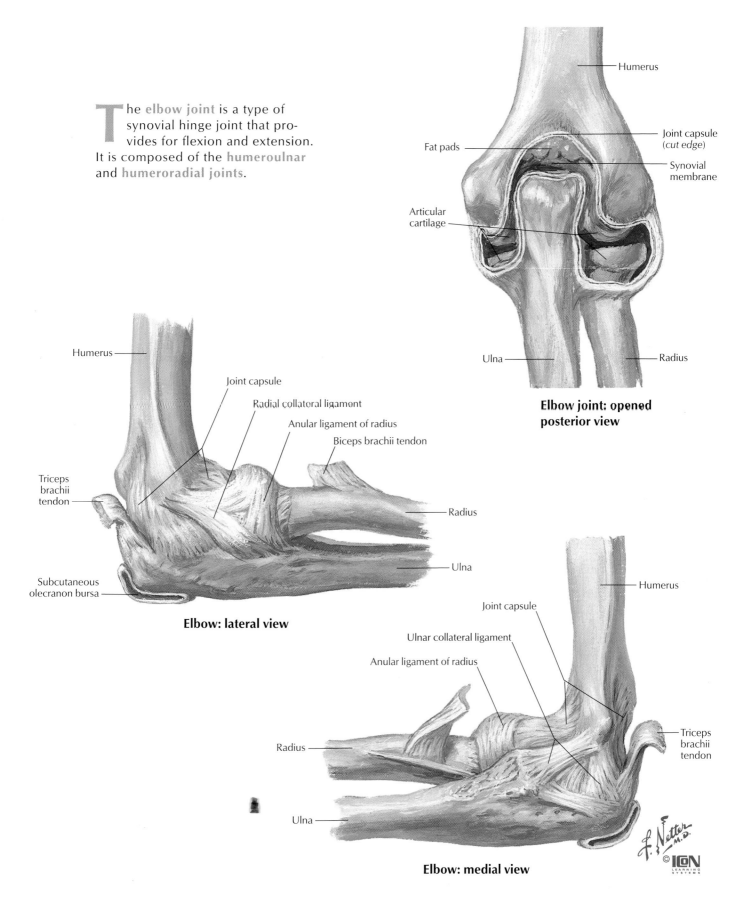

Elbow joint: opened posterior view

Humerus

Joint capsule (cut edge)

Synovial membrane

Fat pads

Articular cartilage

Ulna

Radius

Elbow: lateral view

Humerus

Joint capsule

Radial collateral ligament

Anular ligament of radius

Biceps brachii tendon

Triceps brachii tendon

Radius

Ulna

Subcutaneous olecranon bursa

Elbow: medial view

Joint capsule

Ulnar collateral ligament

Anular ligament of radius

Radius

Humerus

Triceps brachii tendon

Ulna

The Joints & Ligaments
Wrist & Hand

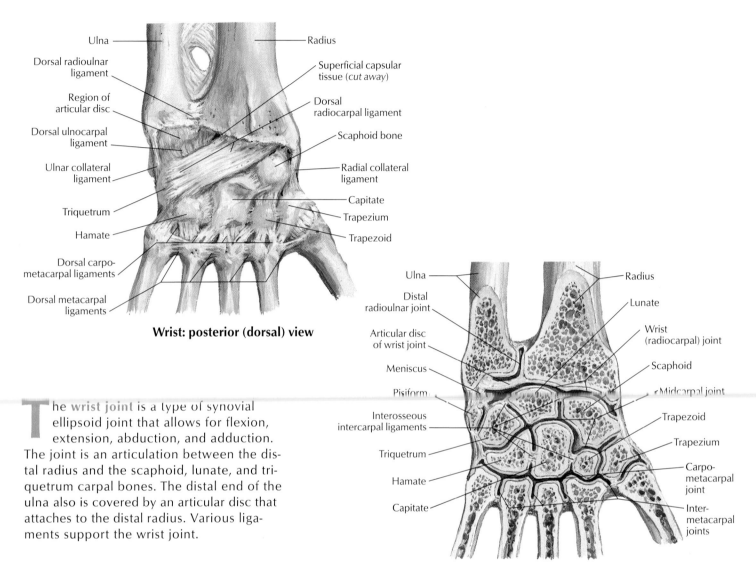

Wrist: posterior (dorsal) view

Ulna —
Dorsal radioulnar ligament —
Region of articular disc —
Dorsal ulnocarpal ligament —
Ulnar collateral ligament —
Triquetrum —
Hamate —
Dorsal carpo-metacarpal ligaments —
Dorsal metacarpal ligaments —

Radius —
Superficial capsular tissue (*cut away*) —
Dorsal radiocarpal ligament —
Scaphoid bone —
Radial collateral ligament —
Capitate —
Trapezium —
Trapezoid —

Wrist: coronal section, dorsal view

Ulna —
Distal radioulnar joint —
Articular disc of wrist joint —
Meniscus —
Pisiform —
Interosseous intercarpal ligaments —
Triquetrum —
Hamate —
Capitate —

Radius —
Lunate —
Wrist (radiocarpal) joint —
Scaphoid —
Midcarpal joint —
Trapezoid —
Trapezium —
Carpo-metacarpal joint —
Inter-metacarpal joints —

The **wrist joint** is a type of synovial ellipsoid joint that allows for flexion, extension, abduction, and adduction. The joint is an articulation between the distal radius and the scaphoid, lunate, and triquetrum carpal bones. The distal end of the ulna also is covered by an articular disc that attaches to the distal radius. Various ligaments support the wrist joint.

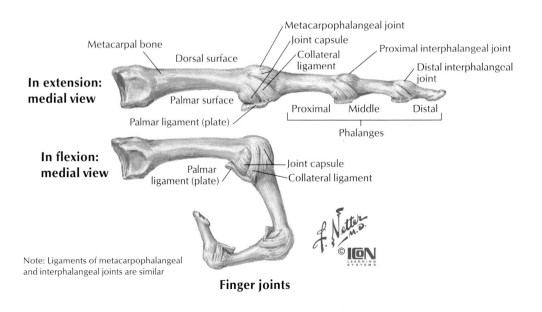

In extension: medial view

Metacarpal bone —
Dorsal surface —
Palmar surface —
Palmar ligament (plate) —

Metacarpophalangeal joint —
Joint capsule —
Collateral ligament —
Proximal interphalangeal joint —
Distal interphalangeal joint —

Proximal Middle Distal
Phalanges

In flexion: medial view

Palmar ligament (plate) —
Joint capsule —
Collateral ligament —

Note: Ligaments of metacarpophalangeal and interphalangeal joints are similar

Finger joints

Ligaments of the Spine

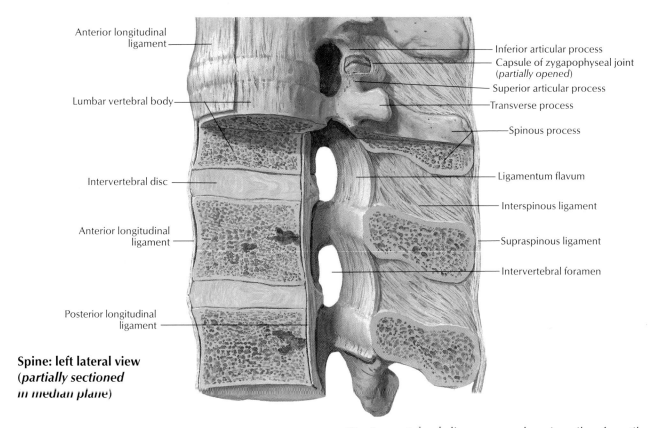

Anterior longitudinal ligament

Lumbar vertebral body

Intervertebral disc

Anterior longitudinal ligament

Posterior longitudinal ligament

Inferior articular process

Capsule of zygapophyseal joint (*partially opened*)

Superior articular process

Transverse process

Spinous process

Ligamentum flavum

Interspinous ligament

Supraspinous ligament

Intervertebral foramen

Spine: left lateral view (*partially sectioned in median plane*)

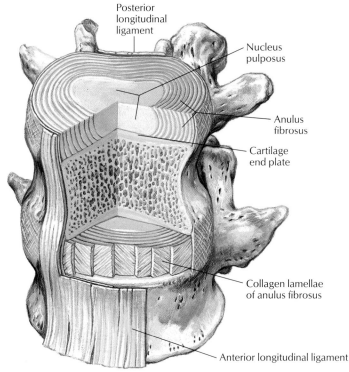

Posterior longitudinal ligament

Nucleus pulposus

Anulus fibrosus

Cartilage end plate

Collagen lamellae of anulus fibrosus

Anterior longitudinal ligament

Intervertebral discs and vertebrae

Intervertebral disc composed of central nuclear zone of collagen and hydrated proteoglycans surrounded by concentric lamellae of collagen fibers

Intravertebral discs are made primarily of cartilage. They separate the bodies of the vertebrae, absorb shock, and prevent the bones from scraping against one another. A normal disc has a gellike center and a tougher outer covering, called the **anulus**, that binds the vertebrae. The squishy interior allows the disc to change shape in response to pressure, allowing the vertebrae to move relative to one another as the spine is flexed and moved. With age, discs become thinner, stiffer, and drier. This diminishes their ability to absorb shock and makes the vertebrae and their joints more vulnerable to damage.

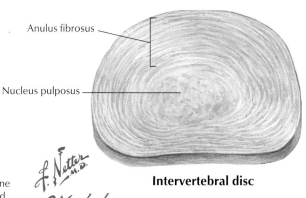

Anulus fibrosus

Nucleus pulposus

Intervertebral disc

Pancreas

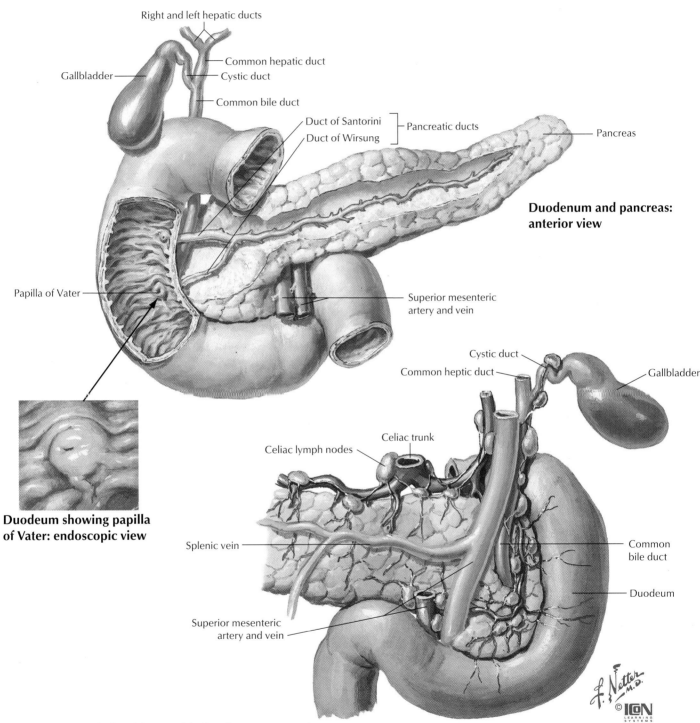

Right and left hepatic ducts

Common hepatic duct

Gallbladder

Cystic duct

Common bile duct

Duct of Santorini — Pancreatic ducts

Duct of Wirsung

Pancreas

Papilla of Vater

Superior mesenteric artery and vein

Duodenum and pancreas: anterior view

Cystic duct

Common heptic duct

Gallbladder

Celiac trunk

Celiac lymph nodes

Splenic vein

Common bile duct

Duodeum

Superior mesenteric artery and vein

Duodeum showing papilla of Vater: endoscopic view

Duodenum and pancreas: posterior view

This elongated gland located behind your stomach and liver fills two roles. First, it produces enzymes that flow into the small intestine to help your body digest proteins, carbohydrates, and fats. Second, it makes hormones that regulate the disposal of nutrients, including sugars. The **islets of Langerhans**, tiny clusters of cells found throughout the pancreas, are responsible for producing these hormones.

Small Intestine

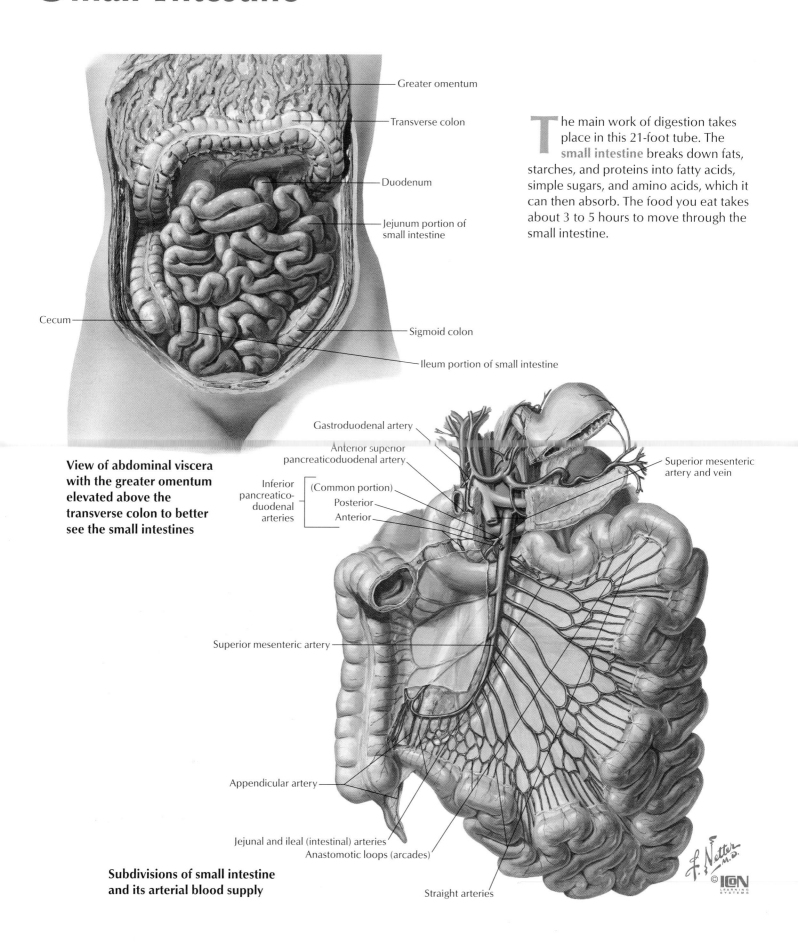

Greater omentum

Transverse colon

Duodenum

Jejunum portion of small intestine

Cecum

Sigmoid colon

Ileum portion of small intestine

The main work of digestion takes place in this 21-foot tube. The **small intestine** breaks down fats, starches, and proteins into fatty acids, simple sugars, and amino acids, which it can then absorb. The food you eat takes about 3 to 5 hours to move through the small intestine.

View of abdominal viscera with the greater omentum elevated above the transverse colon to better see the small intestines

Gastroduodenal artery

Anterior superior pancreaticoduodenal artery

Inferior pancreaticoduodenal arteries

(Common portion)

Posterior

Anterior

Superior mesenteric artery and vein

Superior mesenteric artery

Appendicular artery

Jejunal and ileal (intestinal) arteries

Anastomotic loops (arcades)

Subdivisions of small intestine and its arterial blood supply

Straight arteries

f. Netter M.D.

© ICON
LEARNING SYSTEMS

After the duodenum, the next part of the small intestine is the 8-foot **jejunum**. Here, fats, starches, and proteins are broken down and absorbed. The jejunum surface area is increased by the presence of circular folds covered with **villi** and **microvilli**. This increase in area allows for greater secretion and absorption of ingested food. Isolated **lymph nodules** provide a first-line defense for ingested foreign antigens.

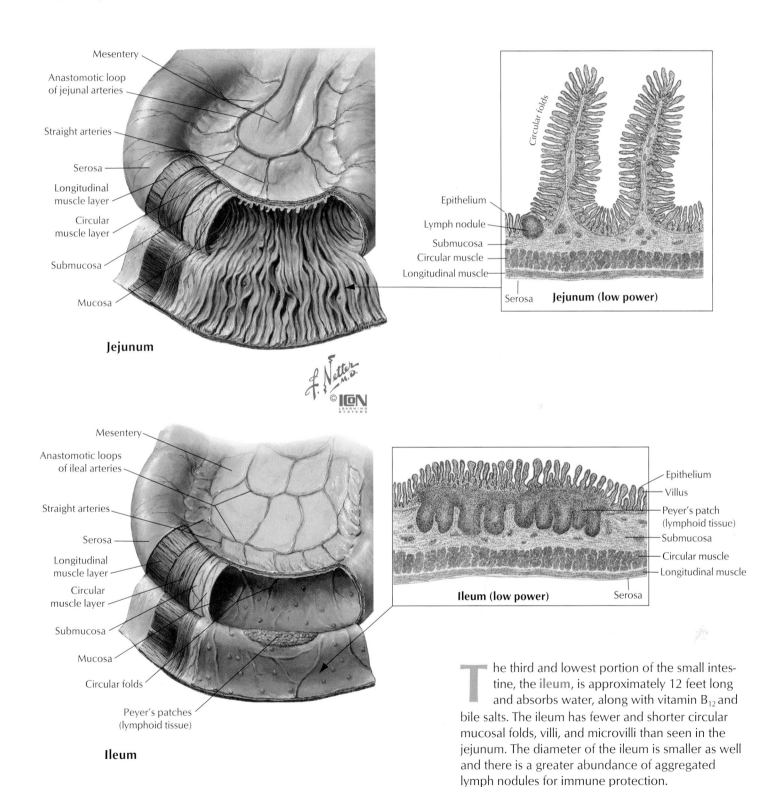

Jejunum

Mesentery
Anastomotic loop of jejunal arteries
Straight arteries
Serosa
Longitudinal muscle layer
Circular muscle layer
Submucosa
Mucosa

Jejunum (low power)

Circular folds
Epithelium
Lymph nodule
Submucosa
Circular muscle
Longitudinal muscle
Serosa

Ileum

Mesentery
Anastomotic loops of ileal arteries
Straight arteries
Serosa
Longitudinal muscle layer
Circular muscle layer
Submucosa
Mucosa
Circular folds
Peyer's patches (lymphoid tissue)

Ileum (low power)

Epithelium
Villus
Peyer's patch (lymphoid tissue)
Submucosa
Circular muscle
Longitudinal muscle
Serosa

The third and lowest portion of the small intestine, the **ileum**, is approximately 12 feet long and absorbs water, along with vitamin B_{12} and bile salts. The ileum has fewer and shorter circular mucosal folds, villi, and microvilli than seen in the jejunum. The diameter of the ileum is smaller as well and there is a greater abundance of aggregated lymph nodules for immune protection.

The Digestive System

Colon

Finally, what is left of the food arrives in the colon, or **large intestine**, a 4-foot-long, muscular tube about the diameter of your fist, where the walls act like a sponge and soak up 80% to 90% of the remaining water not absorbed by the ileum. Once inside the colon, food residue travels up the right side across the **transverse colon**, descends the left side (behind the stomach), passes through the **sigmoid colon** to the **rectum** (behind the left side of the groin), and exits.

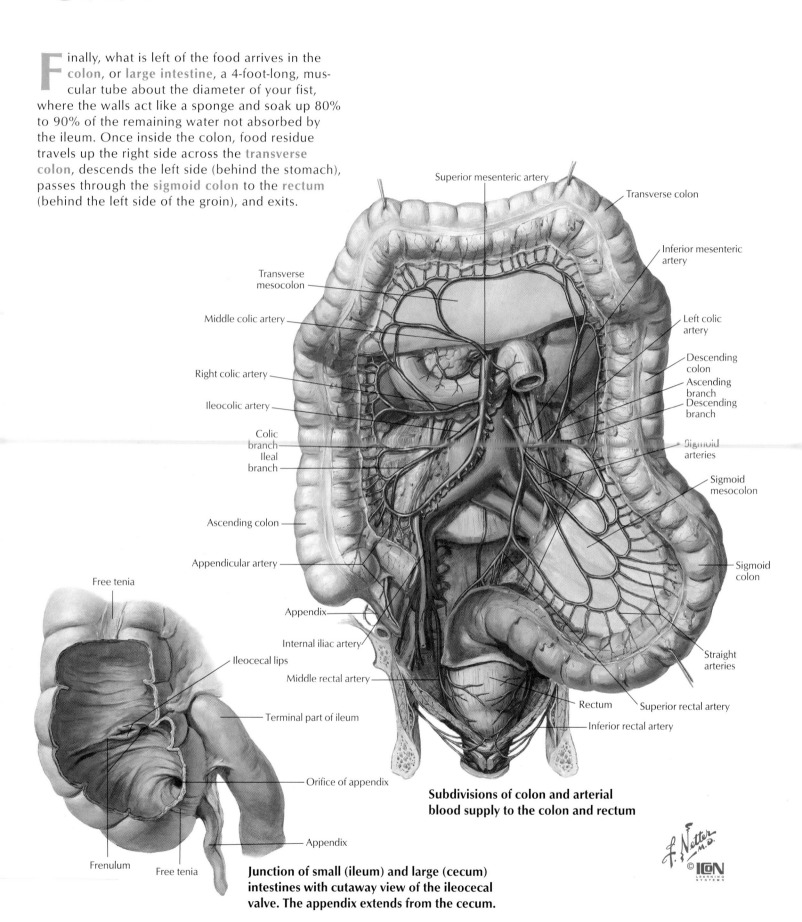

Superior mesenteric artery

Transverse colon

Transverse mesocolon

Middle colic artery

Right colic artery

Ileocolic artery

Colic branch

Ileal branch

Ascending colon

Appendicular artery

Appendix

Internal iliac artery

Middle rectal artery

Inferior mesenteric artery

Left colic artery

Descending colon

Ascending branch

Descending branch

Sigmoid arteries

Sigmoid mesocolon

Sigmoid colon

Straight arteries

Rectum

Superior rectal artery

Inferior rectal artery

Subdivisions of colon and arterial blood supply to the colon and rectum

Free tenia

Ileocecal lips

Terminal part of ileum

Orifice of appendix

Appendix

Frenulum

Free tenia

Junction of small (ileum) and large (cecum) intestines with cutaway view of the ileocecal valve. The appendix extends from the cecum.

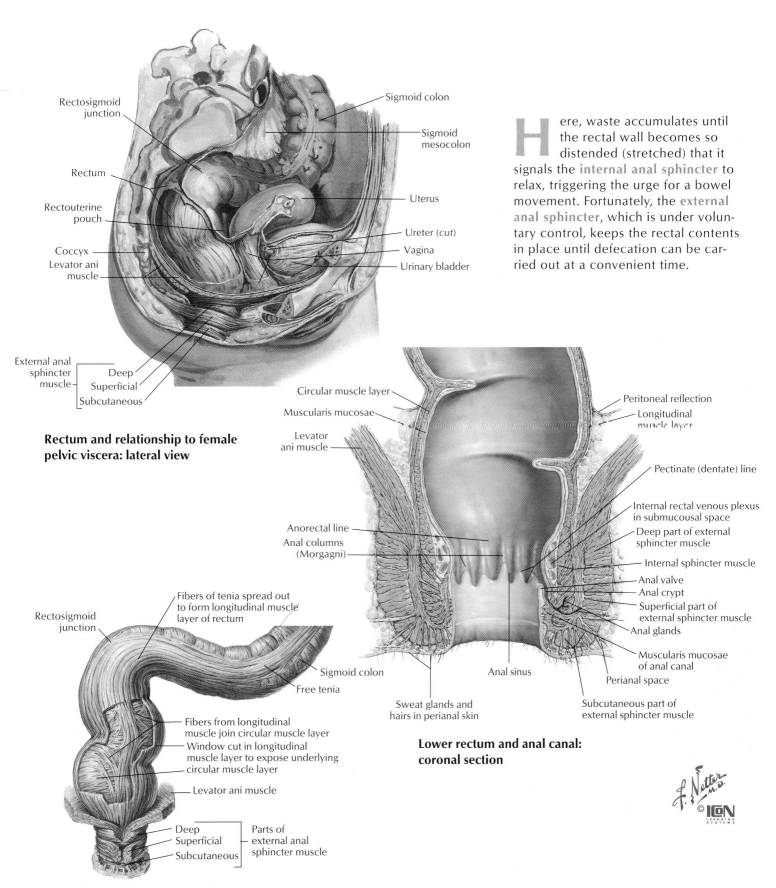

Here, waste accumulates until the rectal wall becomes so distended (stretched) that it signals the internal anal sphincter to relax, triggering the urge for a bowel movement. Fortunately, the external anal sphincter, which is under voluntary control, keeps the rectal contents in place until defecation can be carried out at a convenient time.

Rectum and relationship to female pelvic viscera: lateral view

Labels (top figure):
- Rectosigmoid junction
- Rectum
- Rectouterine pouch
- Coccyx
- Levator ani muscle
- External anal sphincter muscle — Deep / Superficial / Subcutaneous
- Sigmoid colon
- Sigmoid mesocolon
- Uterus
- Ureter (cut)
- Vagina
- Urinary bladder

Lower rectum and anal canal: coronal section

Labels:
- Circular muscle layer
- Muscularis mucosae
- Levator ani muscle
- Anorectal line
- Anal columns (Morgagni)
- Sweat glands and hairs in perianal skin
- Anal sinus
- Peritoneal reflection
- Longitudinal muscle layer
- Pectinate (dentate) line
- Internal rectal venous plexus in submucousal space
- Deep part of external sphincter muscle
- Internal sphincter muscle
- Anal valve
- Anal crypt
- Superficial part of external sphincter muscle
- Anal glands
- Muscularis mucosae of anal canal
- Perianal space
- Subcutaneous part of external sphincter muscle

Rectal musculature and anal sphincter: isolated anterior view

Labels:
- Rectosigmoid junction
- Fibers of tenia spread out to form longitudinal muscle layer of rectum
- Sigmoid colon
- Free tenia
- Fibers from longitudinal muscle join circular muscle layer
- Window cut in longitudinal muscle layer to expose underlying circular muscle layer
- Levator ani muscle
- Deep / Superficial / Subcutaneous — Parts of external anal sphincter muscle

75

The Respiratory System

Our respiratory organs enable us to breathe. A healthy person takes more than 20,000 breaths a day. Each inhalation takes oxygen from the air into the bloodstream and eventually reaches the body's cells. Each exhalation expels **carbon dioxide**, a waste product produced by the body's metabolism. The muscles of the chest wall and the **diaphragm**, the strong muscle that separates the chest from the abdomen, power the work of breathing.

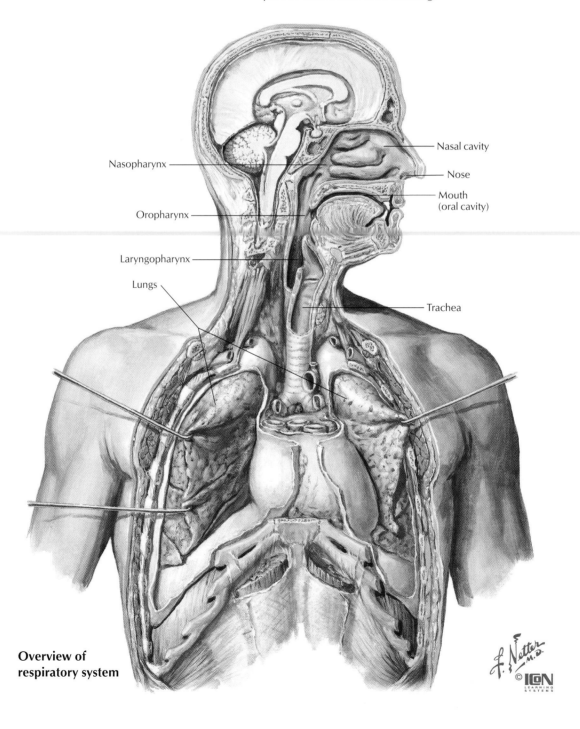

Nasopharynx

Oropharynx

Laryngopharynx

Lungs

Nasal cavity

Nose

Mouth (oral cavity)

Trachea

Overview of respiratory system

Mouth & Throat

W hen a person inhales, air enters the nasal passages and the mouth and then passes through the upper part of the throat (pharynx), the voice box (larynx), and the windpipe (trachea) on its way to the lungs. The epiglottis, a leaf-shaped flap, covers the opening of the larynx during swallowing to keep food and drink from being drawn into the lungs.

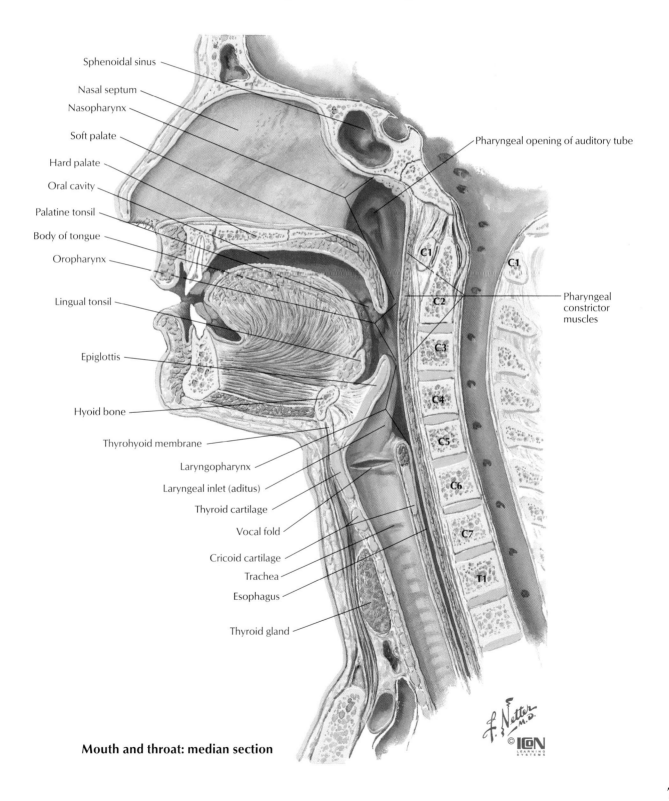

Sphenoidal sinus

Nasal septum

Nasopharynx

Soft palate

Hard palate

Oral cavity

Palatine tonsil

Body of tongue

Oropharynx

Lingual tonsil

Epiglottis

Hyoid bone

Thyrohyoid membrane

Laryngopharynx

Laryngeal inlet (aditus)

Thyroid cartilage

Vocal fold

Cricoid cartilage

Trachea

Esophagus

Thyroid gland

Pharyngeal opening of auditory tube

Pharyngeal constrictor muscles

C1

C2

C3

C4

C5

C6

C7

T1

C1

C1

Mouth and throat: median section

The Respiratory System
Nose & Sinuses

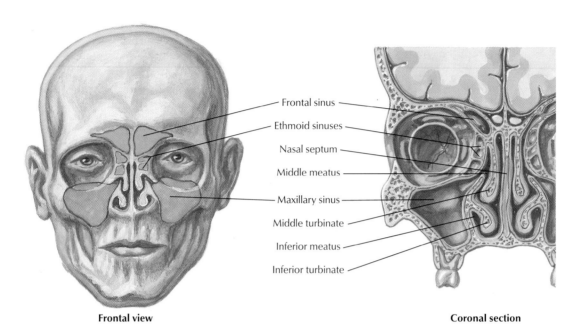

Frontal view

Frontal sinus
Ethmoid sinuses
Nasal septum
Middle meatus
Maxillary sinus
Middle turbinate
Inferior meatus
Inferior turbinate

Coronal section

Anatomy of nasal cavity and sinuses

Your nose is essentially a tent supported by two nasal bones and a midline band of **cartilage**, a firm connective tissue. A thick membrane, the **nasal septum**, runs from the cartilage to separate the nose into two nostrils. As you breathe air into the nose, the hairs that line the nostrils filter any large foreign particles. The air then flows through the **nasal valve**, a narrow passage between the cartilage, septum, and **turbinates**, which are small bones shaped like toy tops. The turbinates are covered in tissue composed of skin cells overlaid with a thick layer of mucus. Your nose also has a rich blood supply. **Sinuses** are the air-filled spaces above, between, behind, and beneath your eyes. The sinuses are lined with a thin membrane that swells and produces **mucus** in response to irritation. Normally, the mucus drains through small openings, known as **ostia**, that connect your sinuses to your nasal passage.

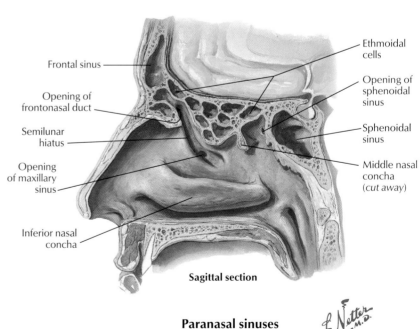

Frontal sinus
Opening of frontonasal duct
Semilunar hiatus
Opening of maxillary sinus
Inferior nasal concha
Ethmoidal cells
Opening of sphenoidal sinus
Sphenoidal sinus
Middle nasal concha (*cut away*)

Sagittal section

Paranasal sinuses

JOHN A. CRAIG—AD
© ICON

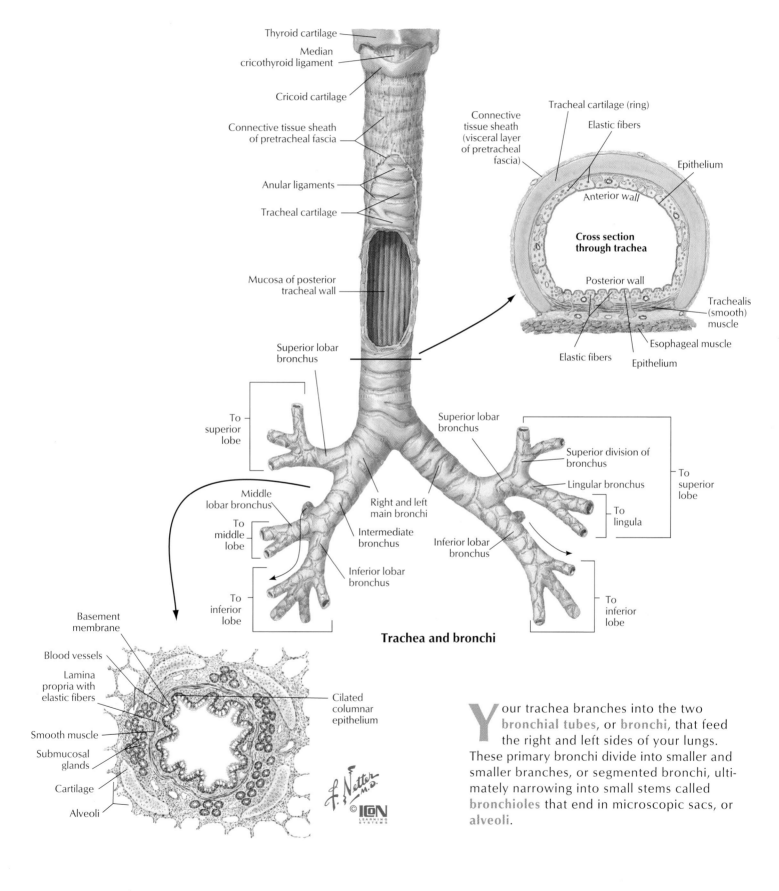

Thyroid cartilage

Median cricothyroid ligament

Cricoid cartilage

Connective tissue sheath of pretracheal fascia

Anular ligaments

Tracheal cartilage

Mucosa of posterior tracheal wall

Superior lobar bronchus

To superior lobe

Middle lobar bronchus

To middle lobe

Right and left main bronchi

Intermediate bronchus

Inferior lobar bronchus

To inferior lobe

Connective tissue sheath (visceral layer of pretracheal fascia)

Tracheal cartilage (ring)

Elastic fibers

Epithelium

Anterior wall

Cross section through trachea

Posterior wall

Trachealis (smooth) muscle

Esophageal muscle

Elastic fibers

Epithelium

Superior lobar bronchus

Superior division of bronchus

Lingular bronchus

To superior lobe

To lingula

Inferior lobar bronchus

To inferior lobe

Trachea and bronchi

Basement membrane

Blood vessels

Lamina propria with elastic fibers

Smooth muscle

Submucosal glands

Cartilage

Alveoli

Cilated columnar epithelium

Your trachea branches into the two bronchial tubes, or bronchi, that feed the right and left sides of your lungs. These primary bronchi divide into smaller and smaller branches, or segmented bronchi, ultimately narrowing into small stems called bronchioles that end in microscopic sacs, or alveoli.

The Respiratory System
Bronchioles & Alveoli

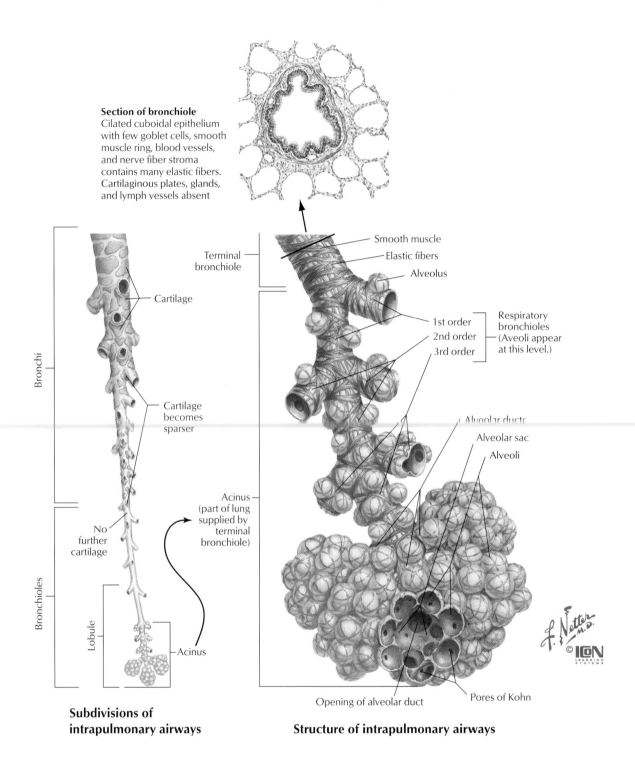

Section of bronchiole
Cilated cuboidal epithelium
with few goblet cells, smooth
muscle ring, blood vessels,
and nerve fiber stroma
contains many elastic fibers.
Cartilaginous plates, glands,
and lymph vessels absent

Smooth muscle

Elastic fibers

Alveolus

Terminal
bronchiole

1st order
2nd order
3rd order

Respiratory
bronchioles
(Aveoli appear
at this level.)

Cartilage

Cartilage
becomes
sparser

Alveolar ducts

Alveolar sac

Alveoli

Bronchi

Acinus
(part of lung
supplied by
terminal
bronchiole)

No
further
cartilage

Bronchioles

Lobule

Acinus

Opening of alveolar duct

Pores of Kohn

**Subdivisions of
intrapulmonary airways**

Structure of intrapulmonary airways

Alveoli are tiny sacs clustered around the bronchioles
in a pattern that resembles bunches of grapes. Each
alveolus is wrapped in a cocoon of lacy capillaries
that bring your blood into contact with the air you pull
into your lungs when you inhale. Although alveoli are
small and round, collectively they create a huge surface
area for the oxygen and carbon dioxide exchange to take
place. If you were to take all your alveoli and spread them
flat, they would cover a tennis court.

Lungs

The two **lungs** are similar in size but can be distinguished anatomically based on their surface features. The right lung has three lobes, demarcated by the **oblique** and **horizontal fissures**, whereas the left lung has only two lobes, demarcated by the oblique fissure. The **hilum** of the lung is situated on its medial aspect and represents the site where structures (vessels, lymphatics, and nerves) enter or leave the lung tissue. Many large structures, such as the major vessels, esophagus, and ribs, lie in proximity to the lungs and leave impressions on the soft lung tissue.

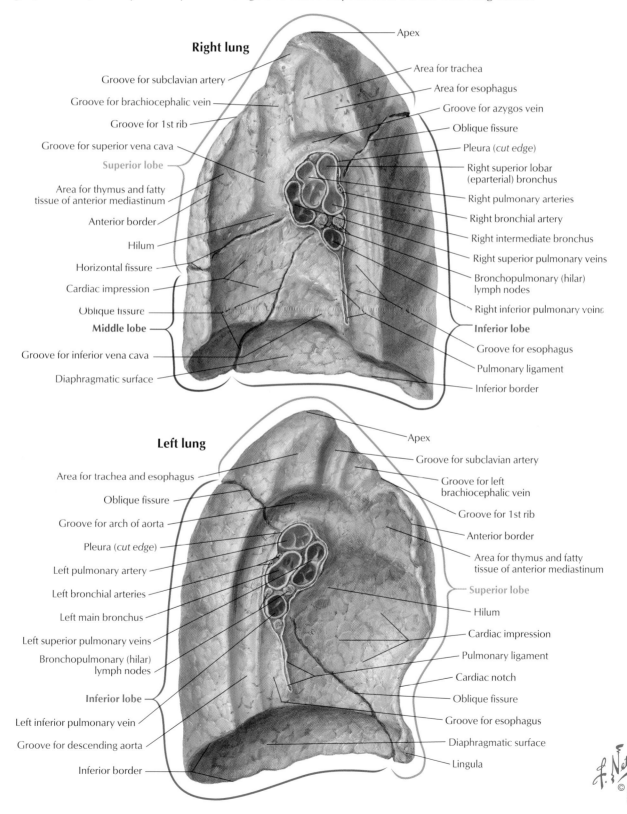

Right lung

- Apex
- Area for trachea
- Area for esophagus
- Groove for subclavian artery
- Groove for brachiocephalic vein
- Groove for 1st rib
- Groove for superior vena cava
- Groove for azygos vein
- Oblique fissure
- Pleura (*cut edge*)
- Superior lobe
- Right superior lobar (eparterial) bronchus
- Area for thymus and fatty tissue of anterior mediastinum
- Right pulmonary arteries
- Anterior border
- Right bronchial artery
- Hilum
- Right intermediate bronchus
- Horizontal fissure
- Right superior pulmonary veins
- Cardiac impression
- Bronchopulmonary (hilar) lymph nodes
- Oblique fissure
- Right inferior pulmonary veins
- Middle lobe
- Inferior lobe
- Groove for inferior vena cava
- Groove for esophagus
- Diaphragmatic surface
- Pulmonary ligament
- Inferior border

Left lung

- Apex
- Groove for subclavian artery
- Area for trachea and esophagus
- Groove for left brachiocephalic vein
- Oblique fissure
- Groove for arch of aorta
- Groove for 1st rib
- Pleura (*cut edge*)
- Anterior border
- Left pulmonary artery
- Area for thymus and fatty tissue of anterior mediastinum
- Left bronchial arteries
- Superior lobe
- Left main bronchus
- Hilum
- Left superior pulmonary veins
- Cardiac impression
- Bronchopulmonary (hilar) lymph nodes
- Pulmonary ligament
- Inferior lobe
- Cardiac notch
- Oblique fissure
- Left inferior pulmonary vein
- Groove for esophagus
- Groove for descending aorta
- Diaphragmatic surface
- Inferior border
- Lingula

The Respiratory System
Gas Exchange

With each intake of breath, your airways dilate and the alveoli expand to admit air, providing oxygen to networks of tiny blood vessels in the alveolar walls. A system of tiny **capillaries** absorbs the oxygen from the alveoli and delivers it throughout the body where it is used by every organ, muscle, and cell. The walls of the bronchi are studded with **mucous glands**, which add moisture to the incoming air. They are lined by tiny hairs called **cilia**, which work with the mucus to trap and expel bacteria and other harmful air-borne particles. When you exhale, the alveoli shrink, forcing exhaled gas—including carbon dioxide—into the bronchioles, back through the bronchi and trachea, and out of your body. Because the bronchi are narrower, it normally takes longer to exhale than to inhale; the narrower your bronchi, the longer it takes to expel air from your lungs. Your brain and nervous system set the pace and rhythm of breathing. In healthy people, rising blood levels of carbon dioxide tell the nervous system to speed up breathing, but in people with advanced lung disease, low oxygen levels can do the same.

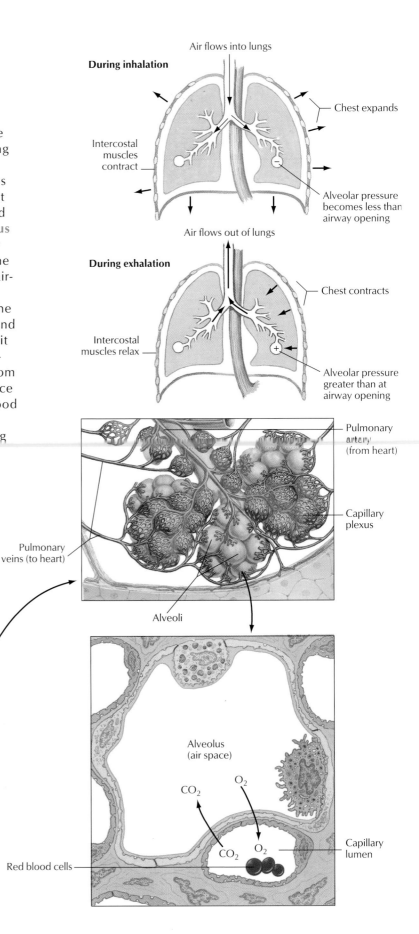

During inhalation

Air flows into lungs

Chest expands

Intercostal muscles contract

Alveolar pressure becomes less than airway opening

During exhalation

Air flows out of lungs

Chest contracts

Intercostal muscles relax

Alveolar pressure greater than at airway opening

Pulmonary artery (from heart)

Capillary plexus

Pulmonary veins (to heart)

Alveoli

Alveolus (air space)

CO_2

O_2

CO_2

O_2

Capillary lumen

Red blood cells

Alveolar capillary

Alveolus

O_2
CO_2

82

The
Cardiovascular System

The **cardiovascular system** includes the **heart**, a powerful muscular pump and a network of arteries, veins, and tiny capillaries that delivers oxygen- and nutrient-rich blood to every cell in your body. A similar network of veins returns blood containing carbon dioxide and other waste materials to your heart and lungs.

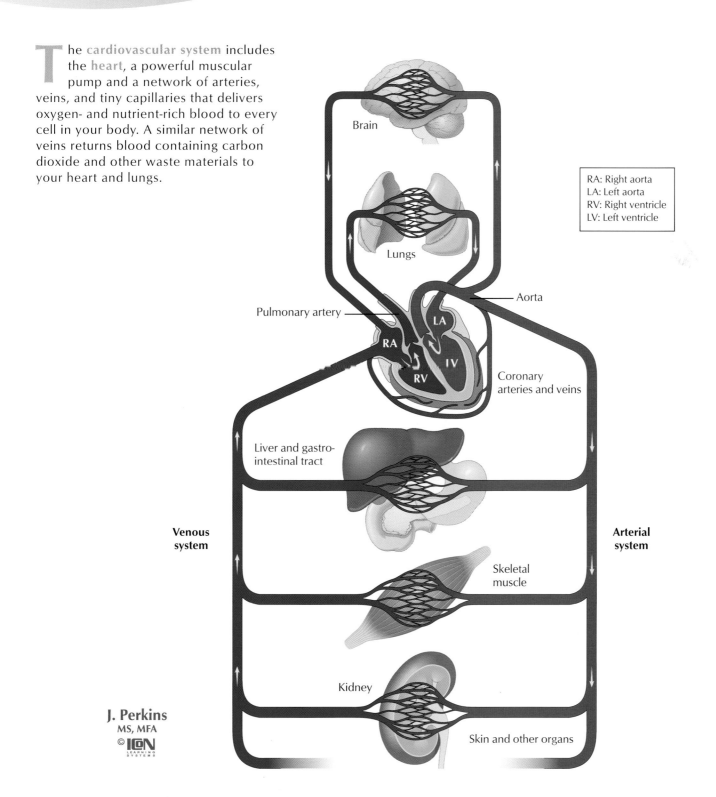

Brain

RA: Right aorta
LA: Left aorta
RV: Right ventricle
LV: Left ventricle

Lungs

Aorta

Pulmonary artery

LA

RA

IV

RV

Coronary
arteries and veins

Liver and gastro-
intestinal tract

Venous
system

Arterial
system

Skeletal
muscle

Kidney

Skin and other organs

J. Perkins
MS, MFA
©ICON
LEARNING
SYSTEMS

Heart

The heart is enclosed within a fibroserous sac called the pericardium. The pericardial cavity is the potential space between the two serous layers of the pericardium and contains a thin film of serous lubricating fluid to reduce friction of the beating heart. The pericardium and heart occupy the middle mediastinum.

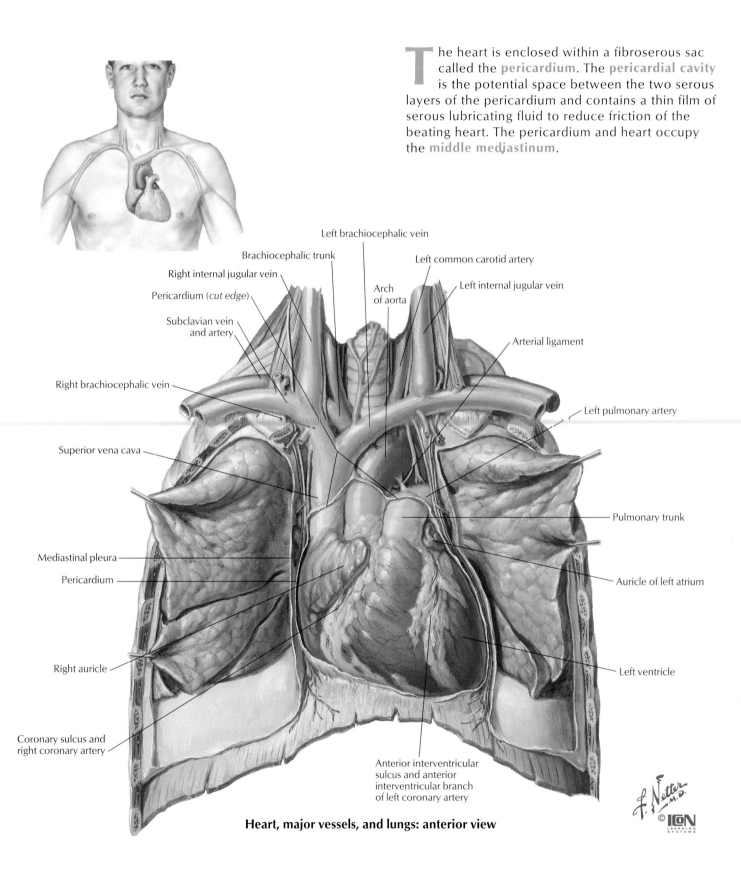

Left brachiocephalic vein

Brachiocephalic trunk

Right internal jugular vein

Left common carotid artery

Pericardium (*cut edge*)

Arch of aorta

Left internal jugular vein

Subclavian vein and artery

Arterial ligament

Right brachiocephalic vein

Left pulmonary artery

Superior vena cava

Pulmonary trunk

Mediastinal pleura

Pericardium

Auricle of left atrium

Right auricle

Left ventricle

Coronary sulcus and right coronary artery

Anterior interventricular sulcus and anterior interventricular branch of left coronary artery

Heart, major vessels, and lungs: anterior view

Your heart is a pump—four pumps, in fact. Blood from all parts of your body collects in veins and flows to the right atrium and the right ventricle, the two pumps that send blood to the lungs. Blood passes through the lungs, where it is cleansed of carbon dioxide and takes up life-sustaining oxygen. The blood then flows to the final two pumps, the left atrium and the left ventricle. The left ventricle collects newly oxygenated blood and, with a muscular push, sends it to all parts of your body. Because the left ventricle has to drive blood to all your body's tissues, it is the largest, strongest, and most important pump.

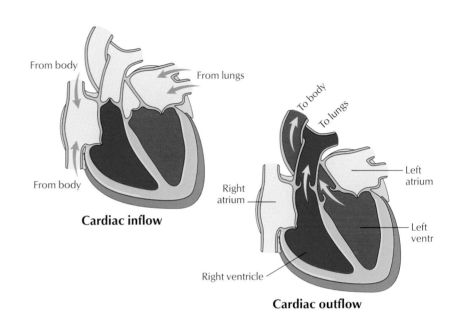

From body

From lungs

From body

Cardiac inflow

To body

To lungs

Right atrium

Left atrium

Left ventr

Right ventricle

Cardiac outflow

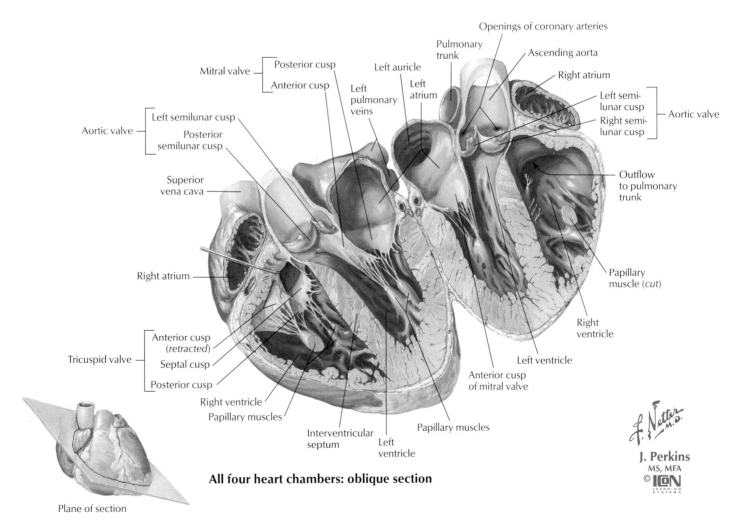

Openings of coronary arteries

Pulmonary trunk

Ascending aorta

Right atrium

Mitral valve — Posterior cusp / Anterior cusp

Left auricle

Left atrium

Left semilunar cusp

Right semilunar cusp

Aortic valve

Left pulmonary veins

Aortic valve — Left semilunar cusp / Posterior semilunar cusp

Superior vena cava

Outflow to pulmonary trunk

Right atrium

Papillary muscle (*cut*)

Right ventricle

Tricuspid valve — Anterior cusp (*retracted*) / Septal cusp / Posterior cusp

Left ventricle

Anterior cusp of mitral valve

Right ventricle

Papillary muscles

Interventricular septum

Left ventricle

Papillary muscles

All four heart chambers: oblique section

Plane of section

J. Netter M.D.

J. Perkins
MS, MFA
© ICON
LEARNING SYSTEMS

Valves

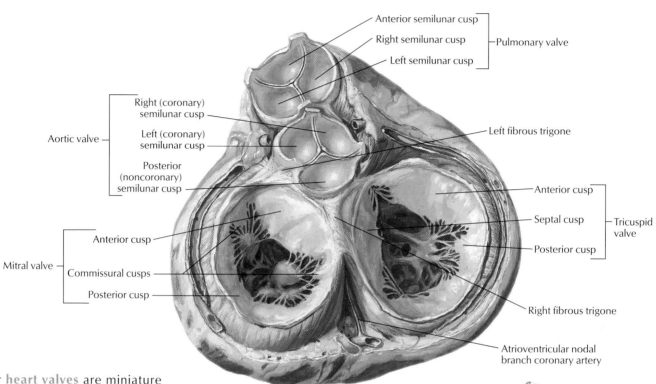

Anterior semilunar cusp
Right semilunar cusp — Pulmonary valve
Left semilunar cusp

Right (coronary)
semilunar cusp
Left (coronary)
Aortic valve — semilunar cusp
Posterior
(noncoronary)
semilunar cusp

Left fibrous trigone

Anterior cusp
Septal cusp — Tricuspid valve
Posterior cusp

Anterior cusp
Mitral valve — Commissural cusps
Posterior cusp

Right fibrous trigone

Atrioventricular nodal
branch coronary artery

**Heart in diastole: viewed from
base with atria removed**

Your **heart valves** are miniature
mechanical marvels. Each has a
set of "flaps" known as **leaflets**
or **cusps**. Normal valves open fully to
let blood flow through, and when
they close, they shut tight. The coor-
dinated opening and closing of these
valves is what keeps blood flowing in
one direction—that is, it ensures that
oxygenated blood flows out to the
body, and back through the lungs
(where it receives a fresh supply of
oxygen) and returns to the left ventricle, where it is once again pumped
to the tissues and organs.

Diastole: Tricuspid and
mitral valves open;
blood flows into
relaxed ventricles

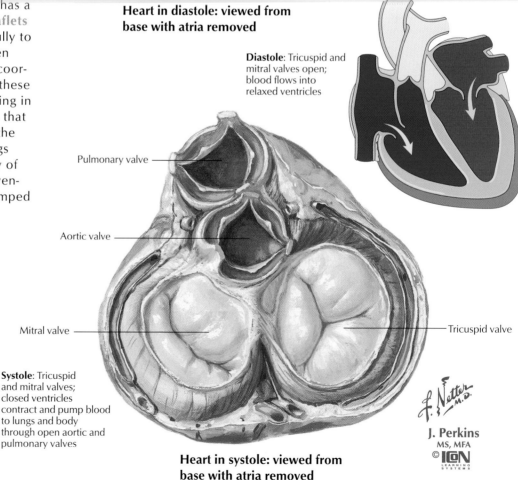

Pulmonary valve

Aortic valve

Mitral valve

Tricuspid valve

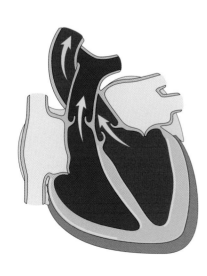

Systole: Tricuspid
and mitral valves;
closed ventricles
contract and pump blood
to lungs and body
through open aortic and
pulmonary valves

**Heart in systole: viewed from
base with atria removed**

J. Perkins
MS, MFA
© ICON
LEARNING
SYSTEMS

Spinal Cord & Peripheral Nerves

The spinal cord runs down the middle of the vertebral column in the vertebral canal, with the spinal nerves exiting the cord via intervertebral openings. All spinal nerves contain motor fibers that innervate skeletal muscle and convey sensory fibers from peripheral receptors back to the spinal cord for higher processing. Some spinal nerves also contain axons of the autonomic nervous system that innervate the heart, smooth muscle (often in the walls of blood vessels and organs), and glands throughout the body.

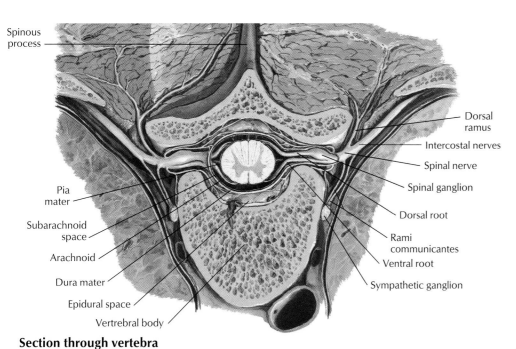

Spinous process
Dorsal ramus
Intercostal nerves
Spinal nerve
Spinal ganglion
Dorsal root
Rami communicantes
Ventral root
Sympathetic ganglion
Pia mater
Subarachnoid space
Arachnoid
Dura mater
Epidural space
Vertrebral body

Section through vertebra

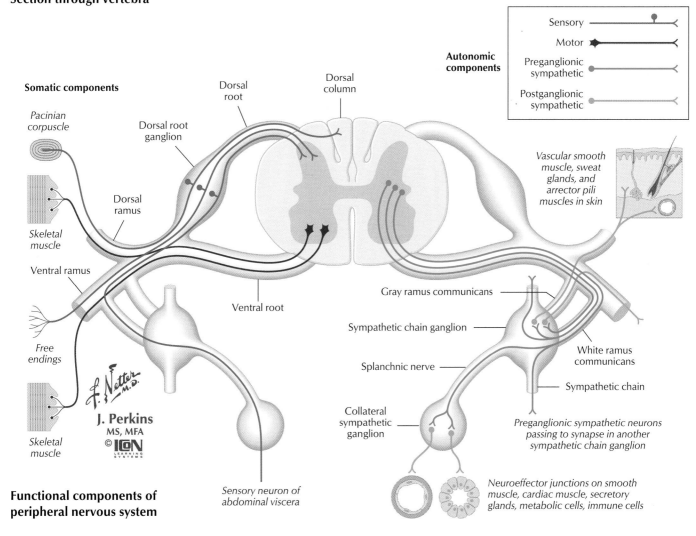

Somatic components

Pacinian corpuscle
Dorsal root
Dorsal root ganglion
Dorsal column
Dorsal ramus
Skeletal muscle
Ventral ramus
Free endings
Skeletal muscle
Ventral root
Sensory neuron of abdominal viscera

Autonomic components

Sensory
Motor
Preganglionic sympathetic
Postganglionic sympathetic

Vascular smooth muscle, sweat glands, and arrector pili muscles in skin

Gray ramus communicans
Sympathetic chain ganglion
Splanchnic nerve
White ramus communicans
Sympathetic chain
Collateral sympathetic ganglion
Preganglionic sympathetic neurons passing to synapse in another sympathetic chain ganglion
Neuroeffector junctions on smooth muscle, cardiac muscle, secretory glands, metabolic cells, immune cells

J. Perkins
MS, MFA
© ICN
LEARNING SYSTEMS

Functional components of peripheral nervous system

The Nervous System
Spinal Nerves & Sensory Dermatomes

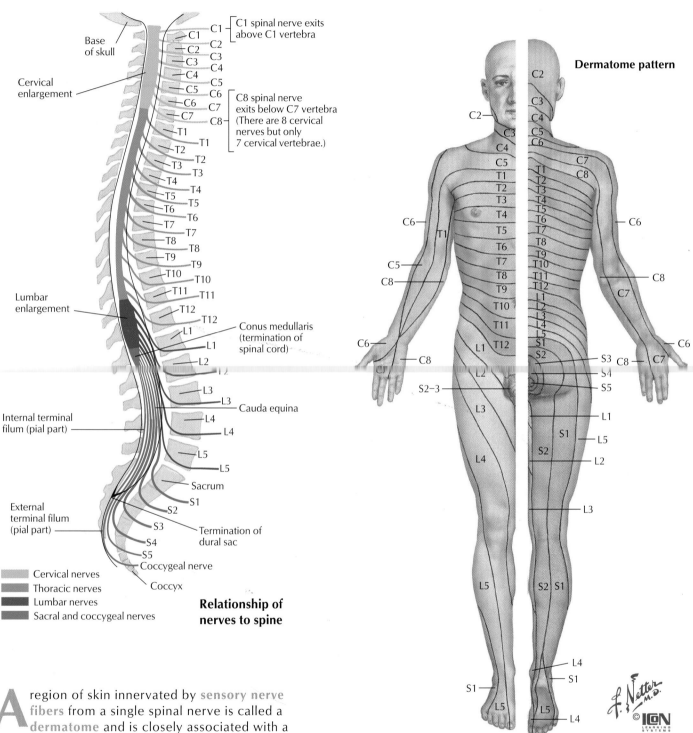

C1 spinal nerve exits above C1 vertebra

C8 spinal nerve exits below C7 vertebra (There are 8 cervical nerves but only 7 cervical vertebrae.)

Base of skull

Cervical enlargement

Lumbar enlargement

Conus medullaris (termination of spinal cord)

Cauda equina

Internal terminal filum (pial part)

External terminal filum (pial part)

Sacrum

Termination of dural sac

Coccygeal nerve

Coccyx

Cervical nerves
Thoracic nerves
Lumbar nerves
Sacral and coccygeal nerves

Relationship of nerves to spine

Dermatome pattern

A region of skin innervated by sensory nerve fibers from a single spinal nerve is called a dermatome and is closely associated with a particular spinal cord level (cervical, thoracic, lumbar, sacral or coccygeal). Physicians can use the segmental dermatome distribution to "map" problems with a nerve or corresponding spinal cord level. Key dermatomes for body region are listed below. For example, skin on your thumb sends sensory axons back to the 6th cervical spinal cord level.

Levels of principal dermatomes

C5	Clavicles
C5, 6, 7	Lateral parts of upper limbs
C8; T1	Medial sides of upper limbs
C6	Thumb
C6, 7, 8	Hand
C8	Ring and little fingers
T4	Level of nipples
T10	Level of umbilicus
T12	Inguinal or groin regions
L1, 2, 3, 4	Anterior and inner surfaces of lower limbs
L4, 5; S1	Foot
L4	Medial side of great toe
L5; S1, 2	Outer and posterior sides of lower limbs
S1	Lateral margin of foot and little toe
S2, 3, 4	Perineum

Autonomic Nervous System

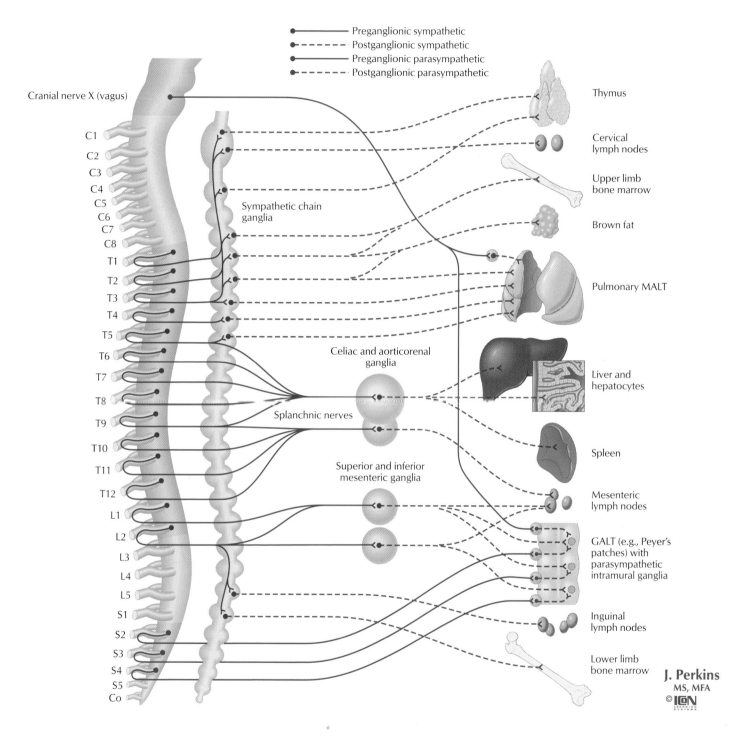

Preganglionic sympathetic
Postganglionic sympathetic
Preganglionic parasympathetic
Postganglionic parasympathetic

Cranial nerve X (vagus)

C1
C2
C3
C4
C5
C6
C7
C8
T1
T2
T3
T4
T5
T6
T7
T8
T9
T10
T11
T12
L1
L2
L3
L4
L5
S1
S2
S3
S4
S5
Co

Sympathetic chain ganglia

Celiac and aorticorenal ganglia

Splanchnic nerves

Superior and inferior mesenteric ganglia

Thymus

Cervical lymph nodes

Upper limb bone marrow

Brown fat

Pulmonary MALT

Liver and hepatocytes

Spleen

Mesenteric lymph nodes

GALT (e.g., Peyer's patches) with parasympathetic intramural ganglia

Inguinal lymph nodes

Lower limb bone marrow

J. Perkins
MS, MFA
©ICON

The autonomic nervous system manages the involuntary activities of smooth muscles, the muscle fibers of internal organs, including the intestines, sweat glands, airways, heart, and blood vessels. The autonomic nervous system is divided into two parts: the sympathetic and the parasympathetic nervous systems. The sympathetic nervous system prepares the body for action by quickening your heart rate and breathing, whereas the parasympathetic nervous system does the opposite. The sympathetic nervous system rules during times of stress or fear. The parasympathetic governs during feeding and sexual arousal.

Facial Nerve

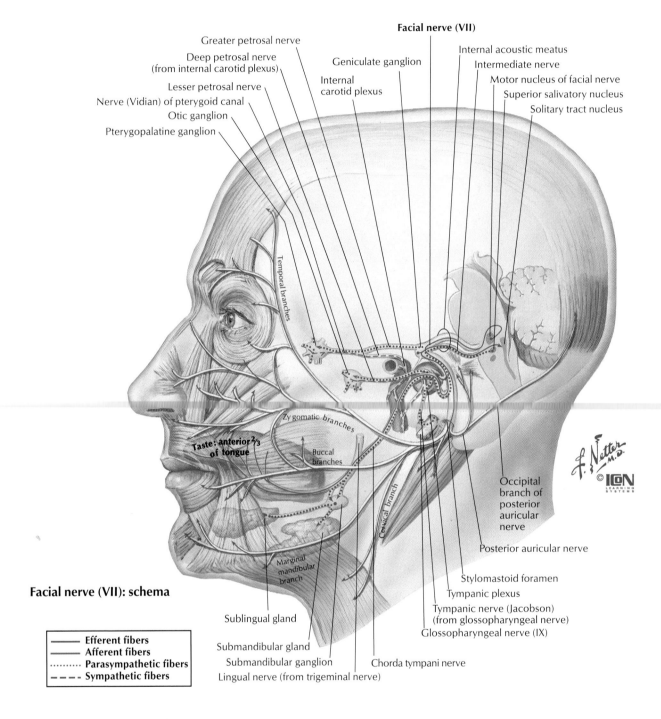

Facial nerve (VII)

Greater petrosal nerve

Deep petrosal nerve
(from internal carotid plexus)

Lesser petrosal nerve

Nerve (Vidian) of pterygoid canal

Otic ganglion

Pterygopalatine ganglion

Geniculate ganglion

Internal
carotid plexus

Internal acoustic meatus

Intermediate nerve

Motor nucleus of facial nerve

Superior salivatory nucleus

Solitary tract nucleus

Temporal branches

Zygomatic branches

Taste: anterior ⅔ of tongue

Buccal branches

Cervical branch

Occipital branch of posterior auricular nerve

Posterior auricular nerve

Marginal mandibular branch

Stylomastoid foramen

Tympanic plexus

Tympanic nerve (Jacobson)
(from glossopharyngeal nerve)

Glossopharyngeal nerve (IX)

Sublingual gland

Submandibular gland

Submandibular ganglion

Lingual nerve (from trigeminal nerve)

Chorda tympani nerve

Facial nerve (VII): schema

——	**Efferent fibers**
——	**Afferent fibers**
··········	**Parasympathetic fibers**
– – – –	**Sympathetic fibers**

The **facial nerve** is the 7th cranial nerve (there are 12 cranial nerves) and it is the great **motor nerve** of the head. It innervates all the muscles of facial expression and several other muscles. However, it also carries autonomic nerve fibers to the mucous glands of the nose and paranasal sinuses, the lacrimal gland (tear production), and two of the three salivary glands (submandibular and sublingual glands) in the oral cavity. Finally, it conveys the sense of taste from the anterior two thirds of the tongue.

Cervical Plexus

Cervical plexus in situ

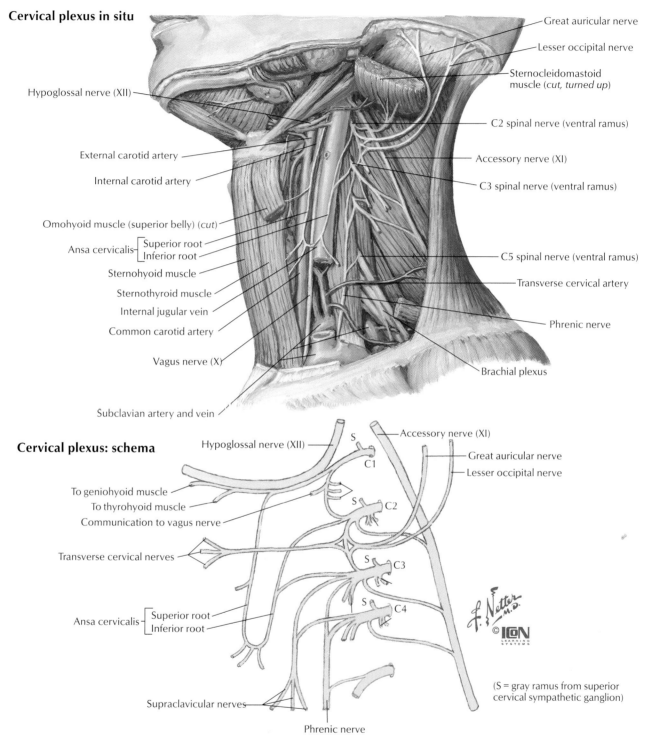

Great auricular nerve

Lesser occipital nerve

Sternocleidomastoid muscle (*cut, turned up*)

Hypoglossal nerve (XII)

C2 spinal nerve (ventral ramus)

External carotid artery

Accessory nerve (XI)

Internal carotid artery

C3 spinal nerve (ventral ramus)

Omohyoid muscle (superior belly) (*cut*)

Ansa cervicalis — Superior root / Inferior root

Sternohyoid muscle

C5 spinal nerve (ventral ramus)

Transverse cervical artery

Sternothyroid muscle

Internal jugular vein

Phrenic nerve

Common carotid artery

Vagus nerve (X)

Brachial plexus

Subclavian artery and vein

Cervical plexus: schema

Hypoglossal nerve (XII)

Accessory nerve (XI)

Great auricular nerve

Lesser occipital nerve

To geniohyoid muscle

To thyrohyoid muscle

Communication to vagus nerve

Transverse cervical nerves

Ansa cervicalis — Superior root / Inferior root

Supraclavicular nerves

Phrenic nerve

(S = gray ramus from superior cervical sympathetic ganglion)

The **cervical plexus** is formed by nerve roots from the first four cervical nerves that arise from the spinal cord. These nerves blend together into a plexus of nerves that innervate many of the muscles of the neck. Unique to this plexus is the formation of a loop, or **ansa cervicalis**, which gives rise to motor nerves to the "strap" muscles in the front of the neck. Also present in this region is the **spinal accessory nerve**, or 11th cranial nerve, which innervates the sternocleidomastoid and trapezius muscles, and the **hypoglossal**, or 12th cranial nerve, which innervates muscles of the tongue.

Brachial Plexus

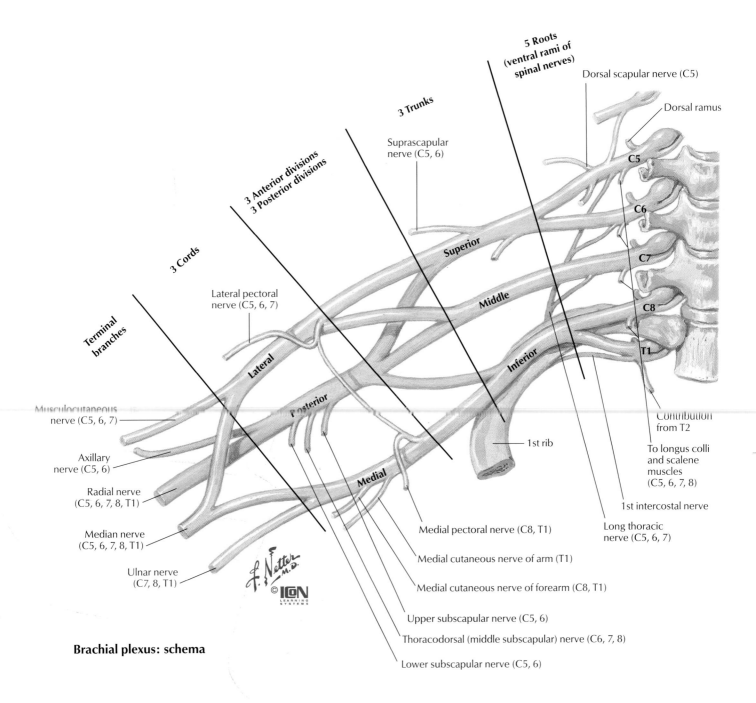

5 Roots (ventral rami of spinal nerves)

Dorsal scapular nerve (C5)

Dorsal ramus

3 Trunks

Suprascapular nerve (C5, 6)

C5

C6

C7

C8

3 Anterior divisions
3 Posterior divisions

Superior

T1

3 Cords

Lateral pectoral nerve (C5, 6, 7)

Middle

Terminal branches

Lateral

Inferior

Contribution from T2

Musculocutaneous nerve (C5, 6, 7)

Posterior

To longus colli and scalene muscles (C5, 6, 7, 8)

Axillary nerve (C5, 6)

1st rib

1st intercostal nerve

Radial nerve (C5, 6, 7, 8, T1)

Medial

Long thoracic nerve (C5, 6, 7)

Median nerve (C5, 6, 7, 8, T1)

Medial pectoral nerve (C8, T1)

Ulnar nerve (C7, 8, T1)

Medial cutaneous nerve of arm (T1)

Medial cutaneous nerve of forearm (C8, T1)

Upper subscapular nerve (C5, 6)

Thoracodorsal (middle subscapular) nerve (C6, 7, 8)

Lower subscapular nerve (C5, 6)

Brachial plexus: schema

Nerves that innervate most of the shoulder muscles and all of the muscles of the upper limb arise from the **brachial plexus**. This plexus is formed by nerve roots from the 5th cervical level to the 1st thoracic level of the spinal cord (C5-T1). Descriptively, the plexus is divided into its five formative *roots*, three *trunks*, six *divisions*, three *cords*, and finally its five *terminal* branches. These five branches pass to the shoulder (axillary nerve) or into the upper limb (radial, musculocutaneous, median, and ulnar nerves) to innervate skeletal muscle.

Median Nerve Distribution

Note: Only muscles innervated by median nerve shown

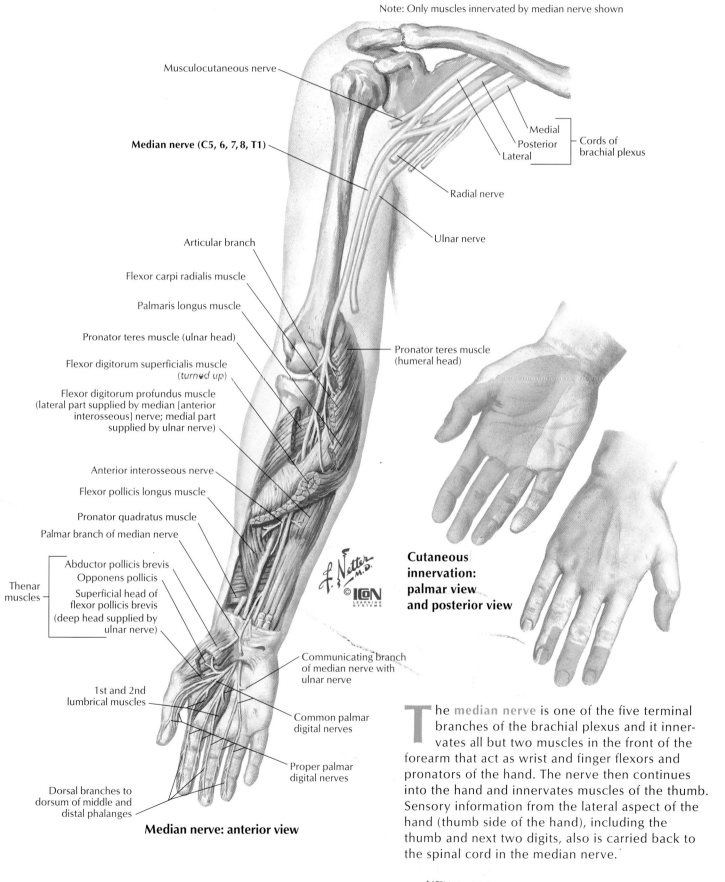

Musculocutaneous nerve

Median nerve (C5, 6, 7, 8, T1)

Medial
Posterior — Cords of
Lateral   brachial plexus

Radial nerve

Ulnar nerve

Articular branch

Flexor carpi radialis muscle

Palmaris longus muscle

Pronator teres muscle (ulnar head)

Flexor digitorum superficialis muscle
(turned up)

Flexor digitorum profundus muscle
(lateral part supplied by median [anterior
interosseous] nerve; medial part
supplied by ulnar nerve)

Pronator teres muscle
(humeral head)

Anterior interosseous nerve

Flexor pollicis longus muscle

Pronator quadratus muscle

Palmar branch of median nerve

Thenar
muscles

Abductor pollicis brevis

Opponens pollicis

Superficial head of
flexor pollicis brevis
(deep head supplied by
ulnar nerve)

Cutaneous
innervation:
palmar view
and posterior view

Communicating branch
of median nerve with
ulnar nerve

1st and 2nd
lumbrical muscles

Common palmar
digital nerves

Proper palmar
digital nerves

Dorsal branches to
dorsum of middle and
distal phalanges

Median nerve: anterior view

The **median nerve** is one of the five terminal branches of the brachial plexus and it innervates all but two muscles in the front of the forearm that act as wrist and finger flexors and pronators of the hand. The nerve then continues into the hand and innervates muscles of syte thumb. Sensory information from the lateral aspect of the hand (thumb side of the hand), including the thumb and next two digits, also is carried back to the spinal cord in the median nerve.

The Nervous System
Lumbar Plexus

The **lumbar plexus** is formed by nerve roots from the 1st lumbar level of the spinal cord to the 4th cord level (L1-L4). Nerves from L4-L5 then continue down to join those of the sacral plexus to innervate the muscles of the pelvis, **perineum** (area between the thighs), and lower limb. The most important nerves of this plexus are the **femoral nerve** (innervates anterior thigh muscles that extend the knee) and the **obturator nerve** (innervates medial thigh muscles of the groin region).

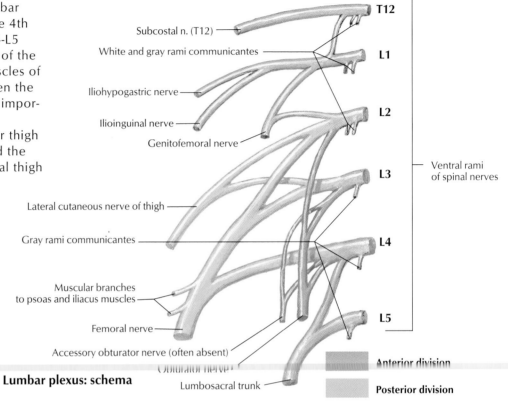

Subcostal n. (T12)
White and gray rami communicantes
Iliohypogastric nerve
Ilioinguinal nerve
Genitofemoral nerve
Lateral cutaneous nerve of thigh
Gray rami communicantes
Muscular branches to psoas and iliacus muscles
Femoral nerve
Accessory obturator nerve (often absent)
Obturator nerve
Lumbosacral trunk

T12
L1
L2
L3
L4
L5

Ventral rami of spinal nerves

Anterior division
Posterior division

Lumbar plexus: schema

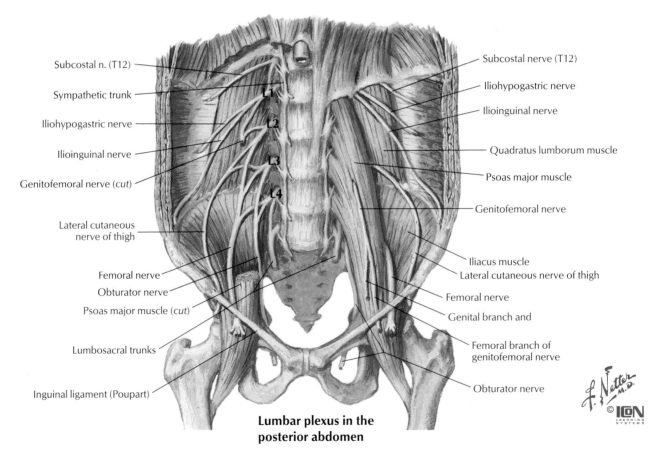

Subcostal n. (T12)
Sympathetic trunk
Iliohypogastric nerve
Ilioinguinal nerve
Genitofemoral nerve (cut)
Lateral cutaneous nerve of thigh
Femoral nerve
Obturator nerve
Psoas major muscle (cut)
Lumbosacral trunks
Inguinal ligament (Poupart)

L1
L2
L3
L4

Subcostal nerve (T12)
Iliohypogastric nerve
Ilioinguinal nerve
Quadratus lumborum muscle
Psoas major muscle
Genitofemoral nerve
Iliacus muscle
Lateral cutaneous nerve of thigh
Femoral nerve
Genital branch and
Femoral branch of genitofemoral nerve
Obturator nerve

Lumbar plexus in the posterior abdomen

Sacral Plexus

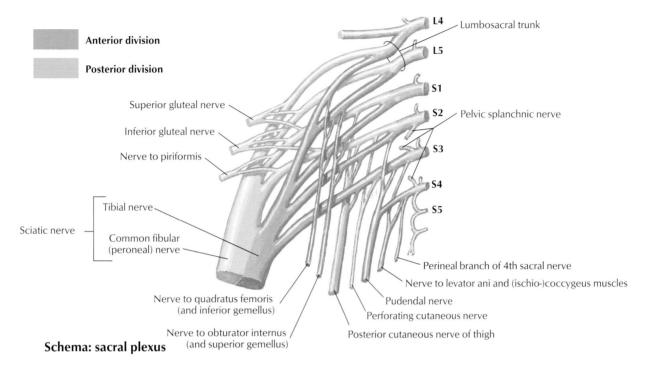

Anterior division

Posterior division

L4 — Lumbosacral trunk

L5

S1

S2 — Pelvic splanchnic nerve

S3

S4

S5

Superior gluteal nerve

Inferior gluteal nerve

Nerve to piriformis

Tibial nerve

Sciatic nerve

Common fibular (peroneal) nerve

Perineal branch of 4th sacral nerve

Nerve to levator ani and (ischio-)coccygeus muscles

Pudendal nerve

Perforating cutaneous nerve

Posterior cutaneous nerve of thigh

Nerve to quadratus femoris (and inferior gemellus)

Nerve to obturator internus (and superior gemellus)

Schema: sacral plexus

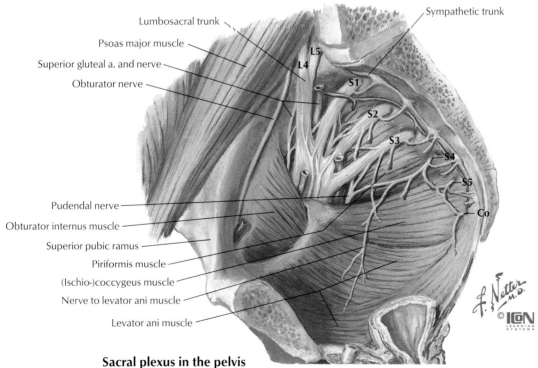

Lumbosacral trunk

Psoas major muscle

Superior gluteal a. and nerve

Obturator nerve

Sympathetic trunk

L5

L4

S1

S2

S3

S4

S5

Co

Pudendal nerve

Obturator internus muscle

Superior pubic ramus

Piriformis muscle

(Ischio-)coccygeus muscle

Nerve to levator ani muscle

Levator ani muscle

Sacral plexus in the pelvis

The sacral plexus, sometimes called the lumbosacral plexus, is formed by nerve root contributions from the L4-S4 spinal cord levels. The plexus lies against the pelvic wall and its major nerves include the gluteal nerves that innervate muscles of the buttock, the pudendal nerve that innervates the perineum (region between the thighs), and the largest nerve in the body, the sciatic nerve. The sciatic nerve innervates all the muscles of the posterior thigh (hamstrings) and all muscles below the level of the knee, including those of the leg and foot.

The Nervous System
Sciatic Nerve

The sciatic nerve is the largest nerve in the body and is really two nerves combined. The larger of the two nerves is the **tibial nerve**, which innervates most of the posterior thigh muscles (hamstrings), the muscles of the posterior leg (calf muscles), and all the muscles of the foot. The smaller nerve component of the sciatic is the **common fibular nerve**, which largely innervates lateral and anterior muscles of the leg. Sensory fibers from the skin cover a large region of the back of the thigh and most of the leg and foot, and are carried in the sciatic nerve and its branches.

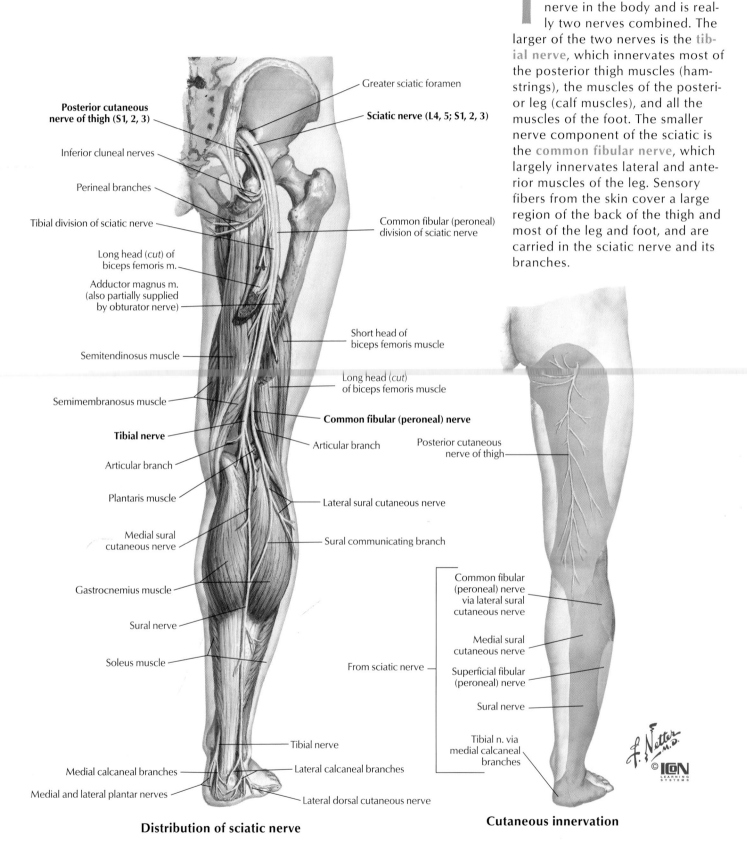

Greater sciatic foramen

Posterior cutaneous nerve of thigh (S1, 2, 3)

Sciatic nerve (L4, 5; S1, 2, 3)

Inferior cluneal nerves

Perineal branches

Tibial division of sciatic nerve

Common fibular (peroneal) division of sciatic nerve

Long head (*cut*) of biceps femoris m.

Adductor magnus m. (also partially supplied by obturator nerve)

Short head of biceps femoris muscle

Semitendinosus muscle

Long head (*cut*) of biceps femoris muscle

Semimembranosus muscle

Common fibular (peroneal) nerve

Tibial nerve

Articular branch

Posterior cutaneous nerve of thigh

Articular branch

Plantaris muscle

Lateral sural cutaneous nerve

Medial sural cutaneous nerve

Sural communicating branch

Gastrocnemius muscle

Common fibular (peroneal) nerve via lateral sural cutaneous nerve

Sural nerve

Medial sural cutaneous nerve

Soleus muscle

Superficial fibular (peroneal) nerve

From sciatic nerve

Sural nerve

Tibial nerve

Tibial n. via medial calcaneal branches

Medial calcaneal branches

Lateral calcaneal branches

Medial and lateral plantar nerves

Lateral dorsal cutaneous nerve

Distribution of sciatic nerve

Cutaneous innervation

The Senses

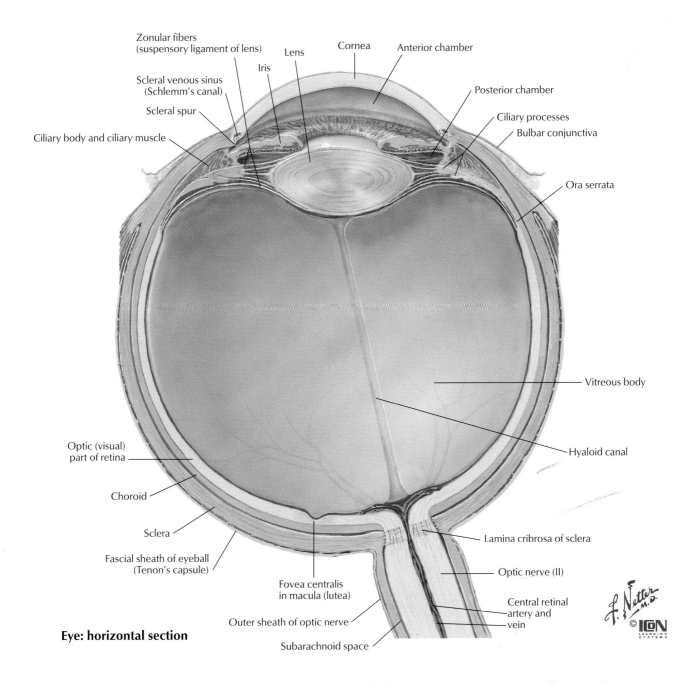

Zonular fibers (suspensory ligament of lens)

Lens

Cornea

Anterior chamber

Iris

Posterior chamber

Scleral venous sinus (Schlemm's canal)

Ciliary processes

Scleral spur

Bulbar conjunctiva

Ciliary body and ciliary muscle

Ora serrata

Vitreous body

Optic (visual) part of retina

Hyaloid canal

Choroid

Sclera

Lamina cribrosa of sclera

Fascial sheath of eyeball (Tenon's capsule)

Optic nerve (II)

Fovea centralis in macula (lutea)

Central retinal artery and vein

Outer sheath of optic nerve

Subarachnoid space

Eye: horizontal section

The outer layer of the eye is made up of the cornea (the transparent dome at the front of your eye) and the white portion, or sclera. The colored iris automatically controls the amount of light entering your eye through the pupil. The lens is directly behind the iris; tiny ligaments attach it to the ciliary muscles on the inner periphery of your eyeball. The aqueous humor is a clear, watery solution that nourishes the lens and helps maintain normal pressure within your eye.

Vision: Orbit and Adnexa of the Eye

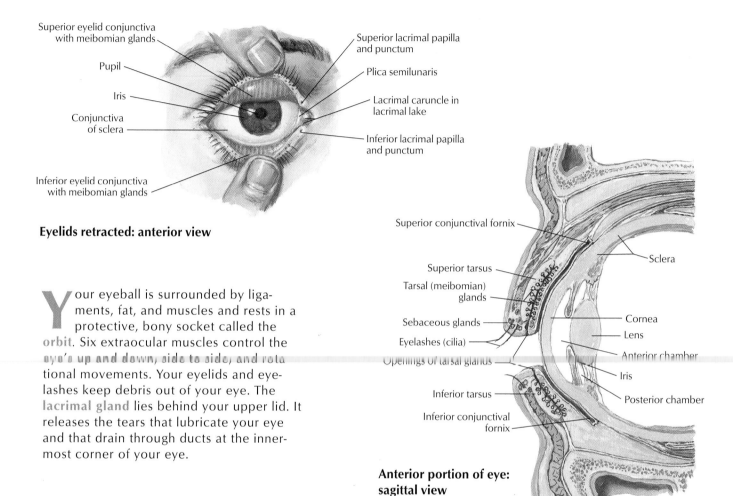

Superior eyelid conjunctiva with meibomian glands

Pupil

Iris

Conjunctiva of sclera

Inferior eyelid conjunctiva with meibomian glands

Superior lacrimal papilla and punctum

Plica semilunaris

Lacrimal caruncle in lacrimal lake

Inferior lacrimal papilla and punctum

Eyelids retracted: anterior view

Superior conjunctival fornix

Superior tarsus

Tarsal (meibomian) glands

Sebaceous glands

Eyelashes (cilia)

Openings of tarsal glands

Inferior tarsus

Inferior conjunctival fornix

Sclera

Cornea

Lens

Anterior chamber

Iris

Posterior chamber

Anterior portion of eye: sagittal view

Your eyeball is surrounded by ligaments, fat, and muscles and rests in a protective, bony socket called the orbit. Six extraocular muscles control the eye's up and down, side to side, and rotational movements. Your eyelids and eyelashes keep debris out of your eye. The lacrimal gland lies behind your upper lid. It releases the tears that lubricate your eye and that drain through ducts at the innermost corner of your eye.

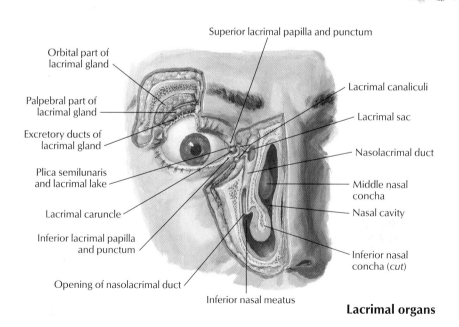

Orbital part of lacrimal gland

Palpebral part of lacrimal gland

Excretory ducts of lacrimal gland

Plica semilunaris and lacrimal lake

Lacrimal caruncle

Inferior lacrimal papilla and punctum

Opening of nasolacrimal duct

Inferior nasal meatus

Superior lacrimal papilla and punctum

Lacrimal canaliculi

Lacrimal sac

Nasolacrimal duct

Middle nasal concha

Nasal cavity

Inferior nasal concha (cut)

Lacrimal organs

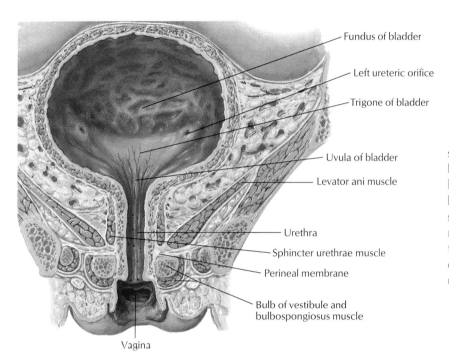

- Fundus of bladder
- Left ureteric orifice
- Trigone of bladder
- Uvula of bladder
- Levator ani muscle
- Urethra
- Sphincter urethrae muscle
- Perineal membrane
- Bulb of vestibule and bulbospongiosus muscle
- Vagina

Female: coronal section

Leading from the bladder to the outside of the body is the **urethra**, a slim and flexible tube surrounded by smooth muscle. At the junction where the bladder meets the urethra (called the **bladder neck**), two sets of muscles help hold urine in. The **internal urethral sphincter** is made up of involuntary muscles that keep a constant pressure on the urethra without conscious effort. The **external**, or outer, **sphincter** muscles are under voluntary control.

In women, the urethra is only 1.5 inches. Estrogen-sensitive cells line the urethra, producing secretions that help create a tight seal. In men, the urethra is about 8 inches, extending from between the scrotum and the rectum to the tip of the penis. Near the bladder, a man's urethra passes through his walnut-sized **prostate gland**. The prostate helps to support the urethra, aiding **continence** (the ability to retain urine), but can also create problems if it becomes enlarged or requires removal or other surgery.

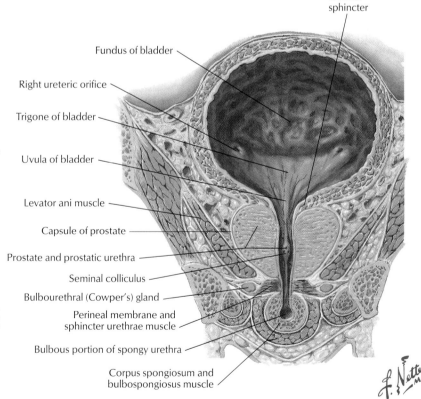

- Internal urethral sphincter
- Fundus of bladder
- Right ureteric orifice
- Trigone of bladder
- Uvula of bladder
- Levator ani muscle
- Capsule of prostate
- Prostate and prostatic urethra
- Seminal colliculus
- Bulbourethral (Cowper's) gland
- Perineal membrane and sphincter urethrae muscle
- Bulbous portion of spongy urethra
- Corpus spongiosum and bulbospongiosus muscle

Male: coronal section

The Urinary System
Pelvic Floor Muscles

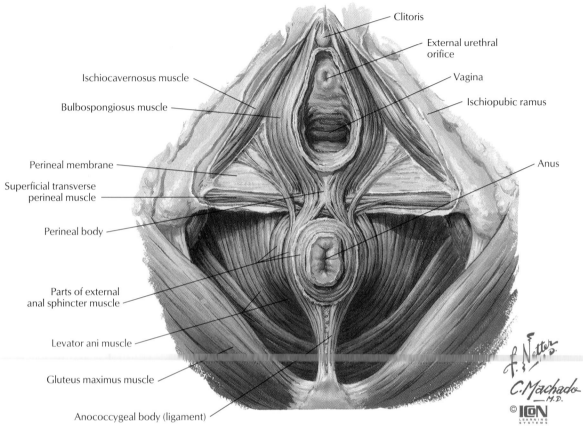

Clitoris

External urethral orifice

Vagina

Ischiopubic ramus

Ischiocavernosus muscle

Bulbospongiosus muscle

Anus

Perineal membrane

Superficial transverse perineal muscle

Perineal body

Parts of external anal sphincter muscle

Levator ani muscle

Gluteus maximus muscle

Anococcygeal body (ligament)

Female: inferior view

Ligaments and the **pelvic floor** support the urinary system, the intestines, and the reproductive organs. The pelvic floor is an important network of muscles that extends from your pubic bone to your tailbone, with openings for the urethra, the anus, and, in women, the vagina. Most of the time, certain pelvic muscles stay contracted to hold the pelvic organs in place against the pull of gravity.

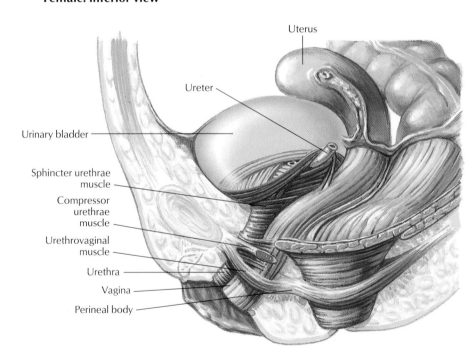

Uterus

Ureter

Urinary bladder

Sphincter urethrae muscle

Compressor urethrae muscle

Urethrovaginal muscle

Urethra

Vagina

Perineal body

Female: lateral view

The Reproductive System

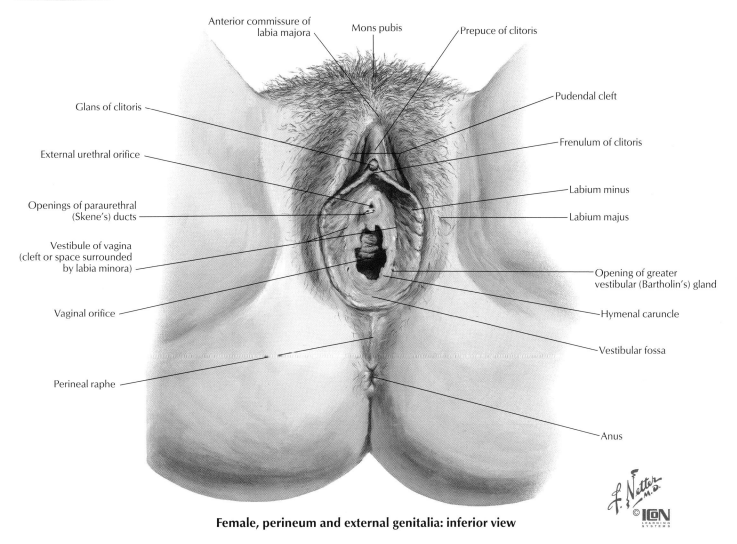

Anterior commissure of labia majora

Mons pubis

Prepuce of clitoris

Glans of clitoris

External urethral orifice

Openings of paraurethral (Skene's) ducts

Vestibule of vagina (cleft or space surrounded by labia minora)

Vaginal orifice

Perineal raphe

Pudendal cleft

Frenulum of clitoris

Labium minus

Labium majus

Opening of greater vestibular (Bartholin's) gland

Hymenal caruncle

Vestibular fossa

Anus

Female, perineum and external genitalia: inferior view

The appearance of a woman's genital organs is as individual as her face or body type. However, certain basic structures are common to all women. The following parts make up the outer genitalia, collectively called the **vulva**:

Mons pubis The fatty mound of tissue that covers the pubic bone. Often called the **mons.**

Outer lips (labia majora) The fleshy folds of skin, fat tissue, and smooth muscle that enclose the vaginal opening. Pubic hair, which may be plentiful or sparse depending on the individual, grows along the outer edges of the labia.

Inner lips (labia minora) A second set of thinner tissue folds, closer to the vaginal opening. Unlike the pubic hair–studded outer lips, the labia minora have a smooth surface and are rich in tiny blood vessels and nerve endings.

Clitoris The most sensitive part of a woman's genital anatomy. This small mound of tissue is located at the point where the upper ends of the labia minora meet above the vaginal opening. It is constructed from the same tissue as the head of a man's penis (the glans). A soft fold of tissue called the **clitoral hood** (or prepuce of clitoris) covers the pea-shaped protrusion.

Perineum The area of anatomy between the thighs, from the pubic bone anteriorly to the coccyx posteriorly.

The Reproductive System
Female Internal Reproductive Organs

Internally, the female reproductive organs include the ovaries (produce the ova, or eggs), the uterine (fallopian) tubes that capture the ovulated egg, the pear-shaped uterus that protects and nourishes the developing fetus, and the musculoelastic distensible vagina that extends from the uterine cervix (neck) to the vaginal orifice (opening).

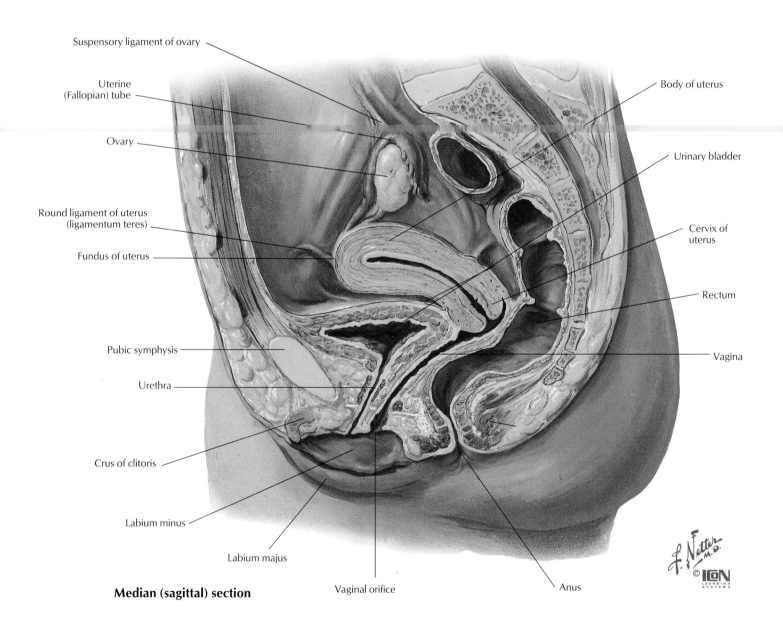

Suspensory ligament of ovary

Uterine (Fallopian) tube

Ovary

Round ligament of uterus (ligamentum teres)

Fundus of uterus

Pubic symphysis

Urethra

Crus of clitoris

Labium minus

Labium majus

Body of uterus

Urinary bladder

Cervix of uterus

Rectum

Vagina

Anus

Vaginal orifice

Median (sagittal) section

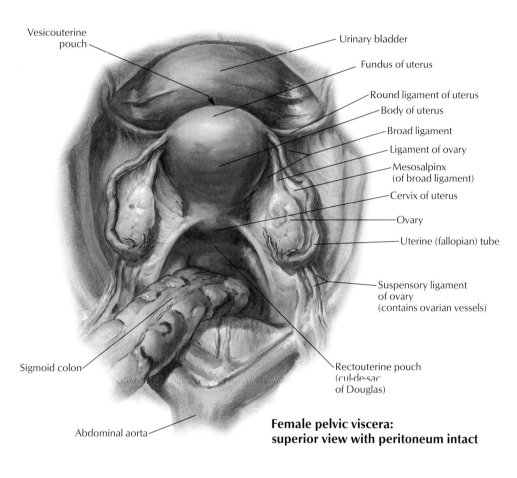

Vesicouterine pouch — Urinary bladder

— Fundus of uterus

— Round ligament of uterus

— Body of uterus

— Broad ligament

— Ligament of ovary

— Mesosalpinx (of broad ligament)

— Cervix of uterus

— Ovary

— Uterine (fallopian) tube

— Suspensory ligament of ovary (contains ovarian vessels)

Sigmoid colon

— Rectouterine pouch (cul-de-sac of Douglas)

Abdominal aorta

**Female pelvic viscera:
superior view with peritoneum intact**

Unseen within a woman's body are the following structures:

Vagina A 3- to 5-inch tube of highly elastic tissue that extends from the vaginal opening to the cervix at the base of the uterus. Just inside the entrance of the vagina is a ridge of muscles. Normally, the vaginal walls rest against one another. During childbirth, however, the vagina stretches wide enough to allow an infant to pass through. The vagina is lined with a layer of cells that secrete fluid to keep the inner surfaces moist. Blood vessels are plentiful beneath the vaginal walls, but most of the nerve endings are clustered in the outer third of the vagina.

Cervix The knoblike tip of the uterus that forms the opening to the uterus from the vagina. Some women find pressure against the cervix enjoyable during intercourse.

Uterus A muscular, fist-sized organ shaped like an upside-down pear. The primary job of the uterus is to harbor a growing fetus during pregnancy. Uterine muscles contract during orgasm, producing a pleasurable sensation.

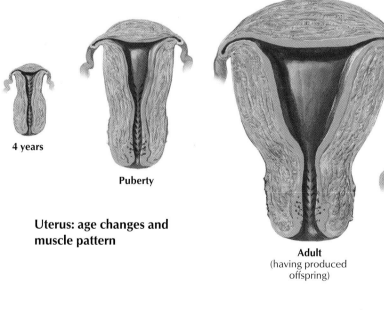

4 years

Puberty

**Uterus: age changes and
muscle pattern**

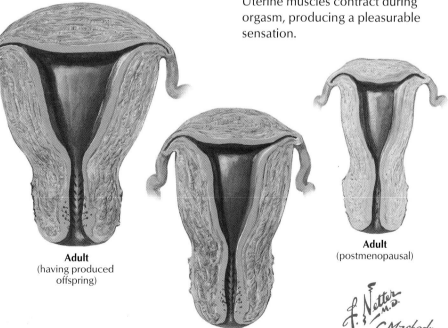

**Adult
(having produced
offspring)**

**Adult
(having never produced
offspring)**

**Adult
(postmenopausal)**

The Reproductive System
Ovary & Ovulation

By puberty, each of this pair of glands contains about 300,000 eggs. Two types of ovarian cysts—the **follicle** and the **corpus luteum**—are a normal part of the reproductive cycle. Hormonal influences cause at least one follicle—a sac containing an egg and fluid—to mature in an **ovary** during each menstrual cycle. During **ovulation**, the follicle ruptures to release the egg. The follicular remnant, or corpus luteum, is a smaller sac containing a thick yellow liquid. It releases **progesterone**, which helps prepare the uterine lining for possible pregnancy.

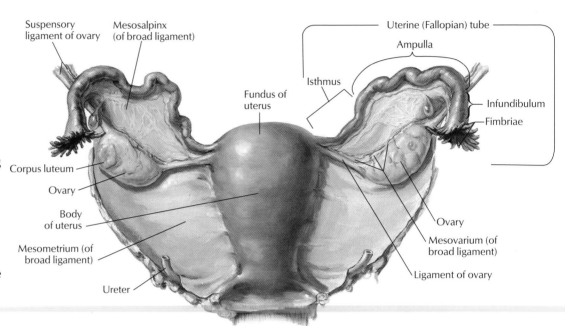

Suspensory ligament of ovary
Mesosalpinx (of broad ligament)
Uterine (Fallopian) tube
Ampulla
Isthmus
Fundus of uterus
Infundibulum
Fimbriae
Corpus luteum
Ovary
Body of uterus
Ovary
Mesovarium (of broad ligament)
Mesometrium (of broad ligament)
Ligament of ovary
Ureter

Uterus and adnexa: posterior view

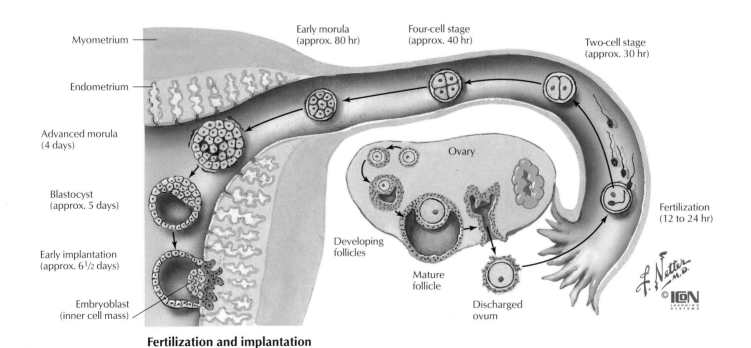

Myometrium
Endometrium
Early morula (approx. 80 hr)
Four-cell stage (approx. 40 hr)
Two-cell stage (approx. 30 hr)
Advanced morula (4 days)
Ovary
Blastocyst (approx. 5 days)
Fertilization (12 to 24 hr)
Early implantation (approx. 6½ days)
Developing follicles
Embryoblast (inner cell mass)
Mature follicle
Discharged ovum

Fertilization and implantation

146

Vagina

T he **vagina** is a distensible fibromuscular tube approximately 3.5 inches in length and is the copulatory organ in women. From its opening, it extends superiorly to join the cervical portion of the uterus. A **fornix**, or troughlike depression, surrounds the site where the vagina and the cervix join. The vagina can stretch significantly during natural childbirth to accommodate the newborn as it is delivered.

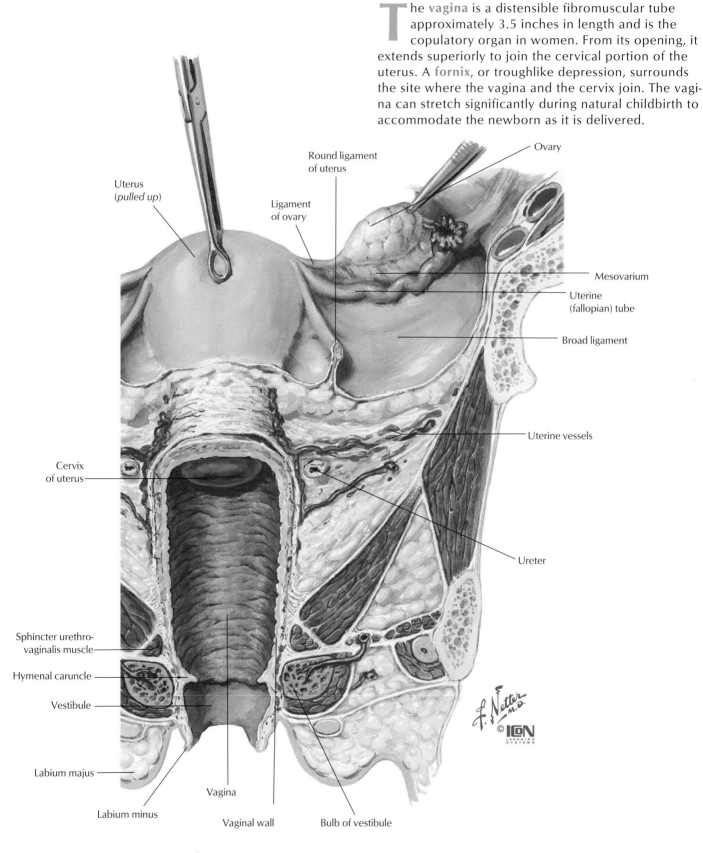

Round ligament of uterus

Ligament of ovary

Uterus (*pulled up*)

Ovary

Mesovarium

Uterine (fallopian) tube

Broad ligament

Uterine vessels

Cervix of uterus

Ureter

Sphincter urethro-vaginalis muscle

Hymenal caruncle

Vestibule

Labium majus

Vagina

Labium minus

Vaginal wall

Bulb of vestibule

Vagina and surrounding structures: frontal view

The Reproductive System
Breast

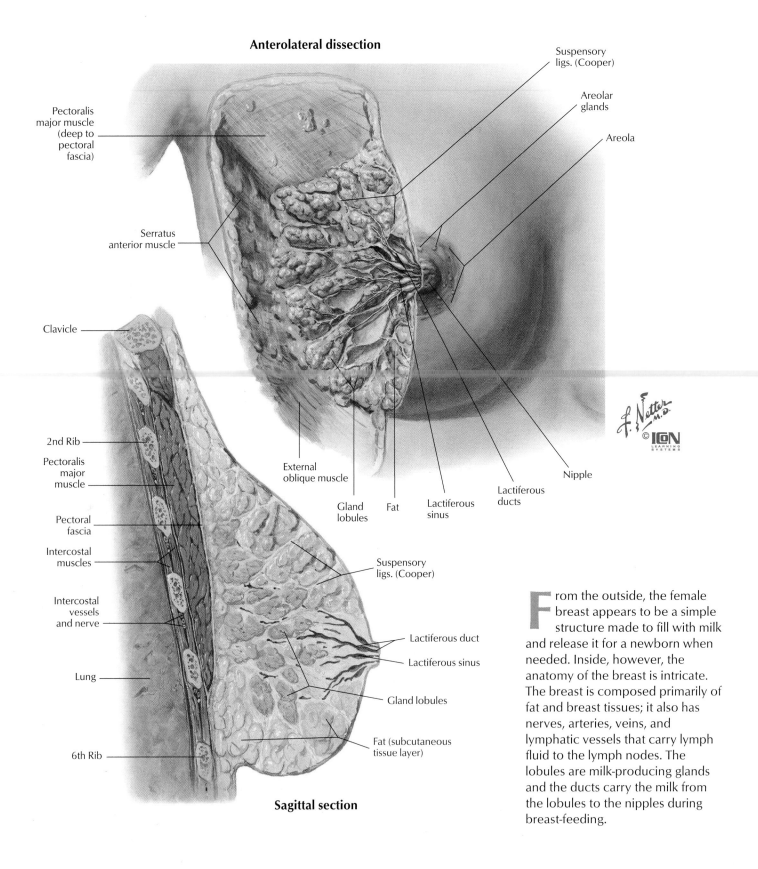

Anterolateral dissection

Pectoralis major muscle (deep to pectoral fascia)

Serratus anterior muscle

Suspensory ligs. (Cooper)

Areolar glands

Areola

Clavicle

2nd Rib

Pectoralis major muscle

Pectoral fascia

Intercostal muscles

Intercostal vessels and nerve

Lung

6th Rib

External oblique muscle

Gland lobules

Fat

Lactiferous sinus

Lactiferous ducts

Nipple

Suspensory ligs. (Cooper)

Lactiferous duct

Lactiferous sinus

Gland lobules

Fat (subcutaneous tissue layer)

Sagittal section

From the outside, the female breast appears to be a simple structure made to fill with milk and release it for a newborn when needed. Inside, however, the anatomy of the breast is intricate. The breast is composed primarily of fat and breast tissues; it also has nerves, arteries, veins, and lymphatic vessels that carry lymph fluid to the lymph nodes. The lobules are milk-producing glands and the ducts carry the milk from the lobules to the nipples during breast-feeding.

Pregnancy—Fetus During Gestation

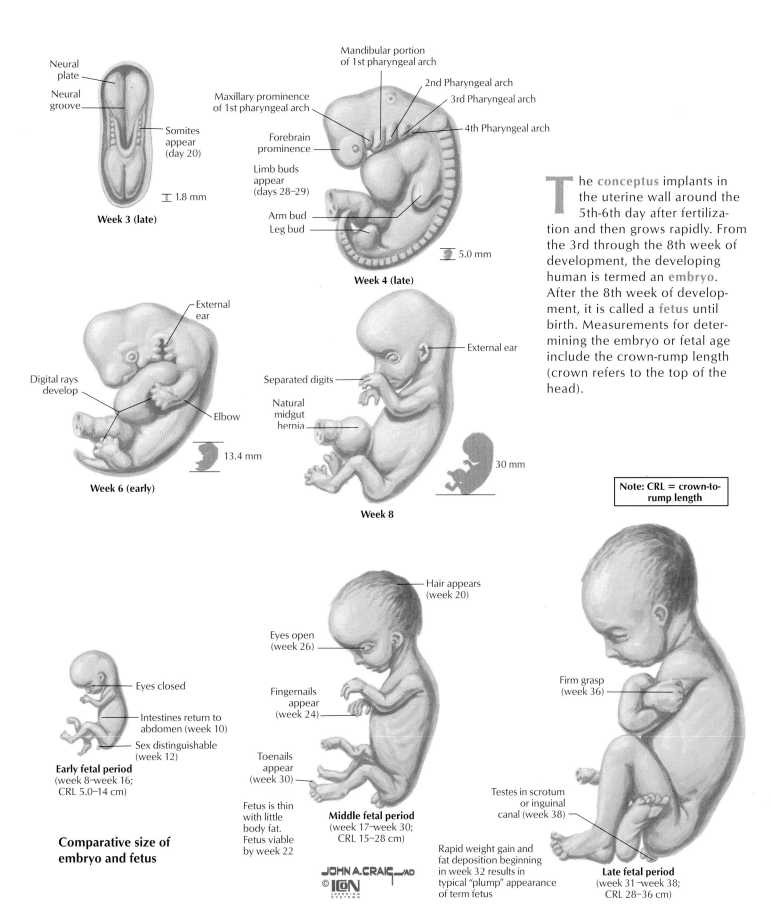

Neural plate

Neural groove

Somites appear (day 20)

1.8 mm

Week 3 (late)

Mandibular portion of 1st pharyngeal arch

2nd Pharyngeal arch

3rd Pharyngeal arch

4th Pharyngeal arch

Maxillary prominence of 1st pharyngeal arch

Forebrain prominence

Limb buds appear (days 28–29)

Arm bud

Leg bud

5.0 mm

Week 4 (late)

The **conceptus** implants in the uterine wall around the 5th-6th day after fertilization and then grows rapidly. From the 3rd through the 8th week of development, the developing human is termed an **embryo**. After the 8th week of development, it is called a **fetus** until birth. Measurements for determining the embryo or fetal age include the crown-rump length (crown refers to the top of the head).

External ear

Digital rays develop

Elbow

13.4 mm

Week 6 (early)

External ear

Separated digits

Natural midgut hernia

30 mm

Week 8

Note: CRL = crown-to-rump length

Eyes closed

Intestines return to abdomen (week 10)

Sex distinguishable (week 12)

Early fetal period
(week 8–week 16; CRL 5.0–14 cm)

Comparative size of embryo and fetus

Hair appears (week 20)

Eyes open (week 26)

Fingernails appear (week 24)

Toenails appear (week 30)

Fetus is thin with little body fat. Fetus viable by week 22

Middle fetal period
(week 17–week 30; CRL 15–28 cm)

Rapid weight gain and fat deposition beginning in week 32 results in typical "plump" appearance of term fetus

Firm grasp (week 36)

Testes in scrotum or inguinal canal (week 38)

Late fetal period
(week 31–week 38; CRL 28–36 cm)

JOHN A.CRAIG—AD
© ICN
LEARNING SYSTEMS

149

The Reproductive System
Pregnancy—The Placenta

The **umbilical cord** is the fetus' lifeline to the **placenta**, which represents a specialized elaboration of embryonic and maternal tissues that allows for the maternal nurturing of the growing embryo and fetus. Nutrients, metabolic gases (oxygen and carbon dioxide), vitamins, steroid hormones, and many other substances cross the fetal-maternal barrier at the placenta. The placenta does not permit certain harmful substances to cross this barrier, but some substances, such as alcohol, many viruses, and some drugs, do cross and can cause potential harm to the fetus.

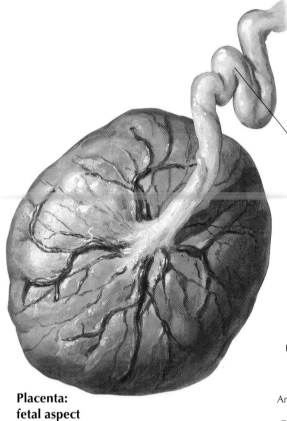

**Placenta:
fetal aspect**

Umbilical cord

Amniochorionic
membrane

Placenta

Full-term fetus within the uterus

Umbilical
arteries

Umbilical vein

Amnion

Syncytiotrophoblast

Chorion

Cytotrophoblast

Cotyledons

Free villi

Connective
tissue septa

Decidua
basalis

Anchoring
villus

Maternal
vessels

Cytotrophoblastic
shell

Development of the placenta: chorionic villi

Fetal versus Postnatal Circulation

Postnatally, blood no longer passes through the placenta but does perfuse the newborn's lungs. Consequently, prenatal shunts that delivered blood to and from the placenta (umbilical vessels) and bypassed the fetal liver (ductus venosus), right ventricle of the heart (foramen ovale), and pulmonary circulation (ductus arteriosus) now close and become ligaments. Thus, the newborn establishes its own pulmonary and systemic circulations independent of the placenta.

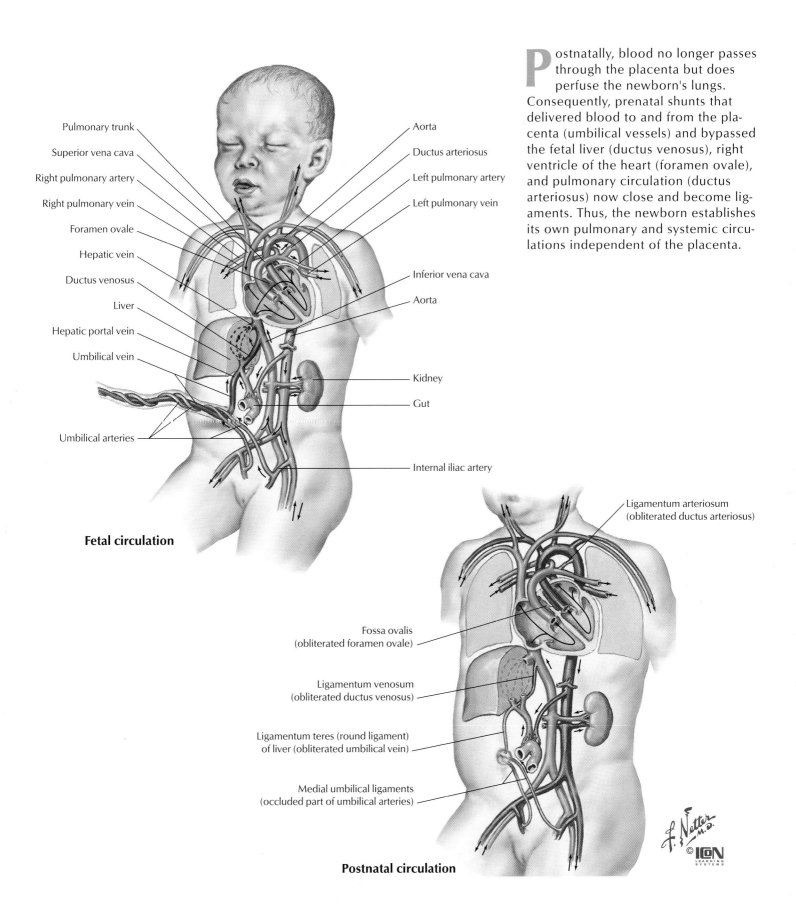

Pulmonary trunk

Superior vena cava

Right pulmonary artery

Right pulmonary vein

Foramen ovale

Hepatic vein

Ductus venosus

Liver

Hepatic portal vein

Umbilical vein

Umbilical arteries

Fetal circulation

Aorta

Ductus arteriosus

Left pulmonary artery

Left pulmonary vein

Inferior vena cava

Aorta

Kidney

Gut

Internal iliac artery

Ligamentum arteriosum (obliterated ductus arteriosus)

Fossa ovalis (obliterated foramen ovale)

Ligamentum venosum (obliterated ductus venosus)

Ligamentum teres (round ligament) of liver (obliterated umbilical vein)

Medial umbilical ligaments (occluded part of umbilical arteries)

Postnatal circulation

151

The Reproductive System
Male Reproductive Organs

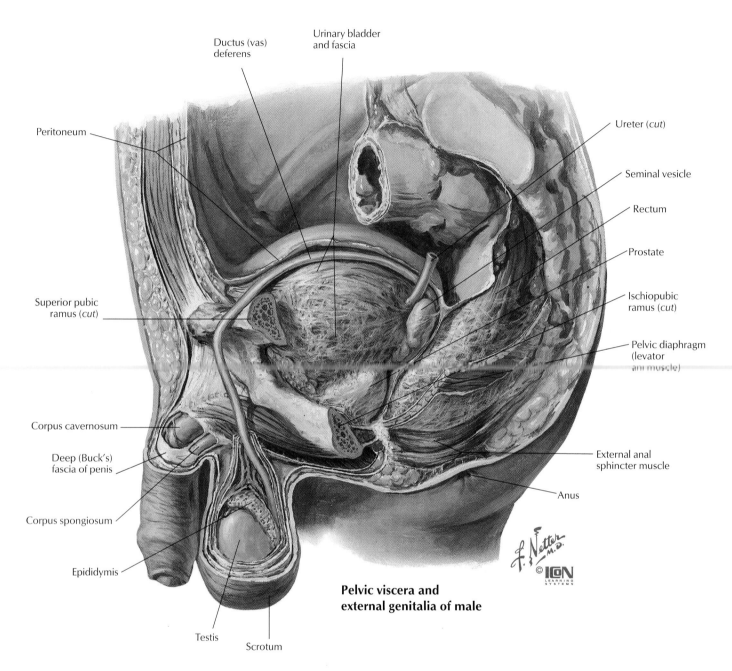

Ductus (vas) deferens

Urinary bladder and fascia

Peritoneum

Ureter (*cut*)

Seminal vesicle

Rectum

Prostate

Superior pubic ramus (*cut*)

Ischiopubic ramus (*cut*)

Pelvic diaphragm (levator ani muscle)

Corpus cavernosum

Deep (Buck's) fascia of penis

External anal sphincter muscle

Corpus spongiosum

Anus

Epididymis

Testis

Scrotum

Pelvic viscera and external genitalia of male

In the male, the **gonads** (**testes**) reside outside the body in the **scrotum**, and the **penis** is used both to expel urine and to transmit **semen** (sperm and seminal fluid) during copulation with the female. Sperm are produced in the testes, mature in the epididymis, are conveyed via the ductus deferens to the ejaculatory duct within the prostate gland (seminal vesicle and prostatic fluids added here), and then enter the male urethra for ejaculation.

Penis

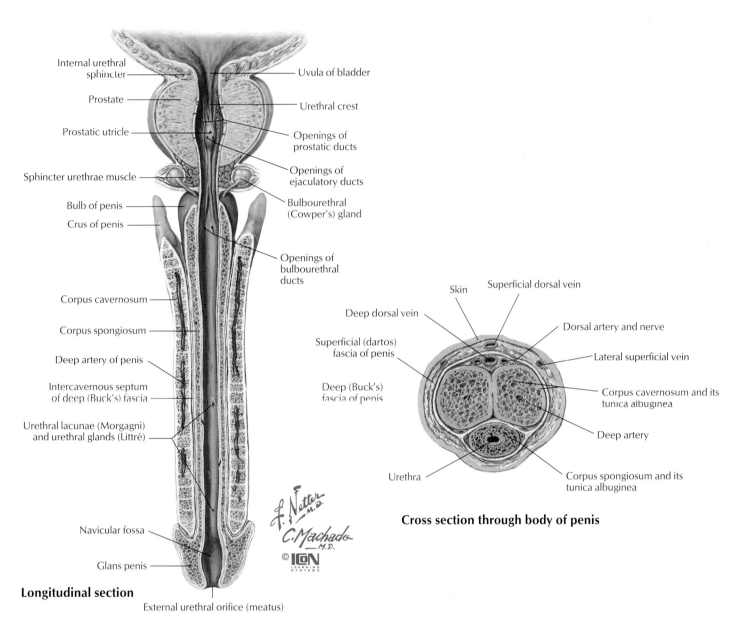

Internal urethral sphincter

Uvula of bladder

Prostate

Urethral crest

Prostatic utricle

Openings of prostatic ducts

Sphincter urethrae muscle

Openings of ejaculatory ducts

Bulb of penis

Crus of penis

Bulbourethral (Cowper's) gland

Openings of bulbourethral ducts

Corpus cavernosum

Corpus spongiosum

Deep artery of penis

Intercavernous septum of deep (Buck's) fascia

Urethral lacunae (Morgagni) and urethral glands (Littré)

Navicular fossa

Glans penis

Longitudinal section

External urethral orifice (meatus)

Skin

Superficial dorsal vein

Deep dorsal vein

Dorsal artery and nerve

Superficial (dartos) fascia of penis

Lateral superficial vein

Deep (Buck's) fascia of penis

Corpus cavernosum and its tunica albuginea

Deep artery

Urethra

Corpus spongiosum and its tunica albuginea

Cross section through body of penis

The penis has the following structures:

Glans The head of the penis. The urethral opening at the tip of the glans allows urine and semen to leave the penis.

Corona The ridge that separates the glans from the shaft. This and the glans are the most sensitive portions of a man's penis.

Shaft The main part of the penis. It houses the corpora cavernosa and the corpus spongiosum.

Corpora cavernosa (erectile bodies) Two flexible cylinders of erectile tissue that run the length of the penis to support erection.

Corpus spongiosum (spongy body) A cylindrical body of erectile tissue that surrounds the urethra and includes the glans.

Central artery The vessel that supplies blood to erectile tissue in the corpora cavernosa.

Urethra A narrow tube that extends the length of the prostate and penis and carries urine and semen out of the body.

153

Prostate

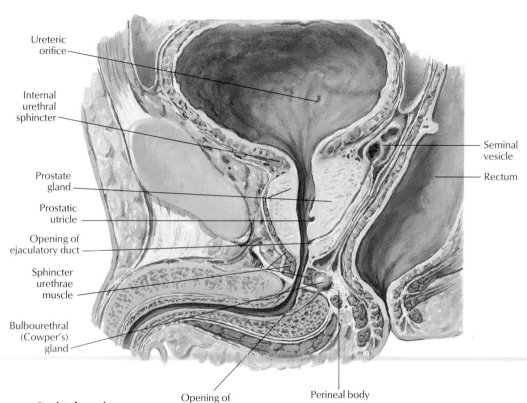

Ureteric orifice

Internal urethral sphincter

Prostate gland

Prostatic utricle

Opening of ejaculatory duct

Sphincter urethrae muscle

Bulbourethral (Cowper's) gland

Seminal vesicle

Rectum

Opening of bulbourethral duct

Perineal body

Sagittal section

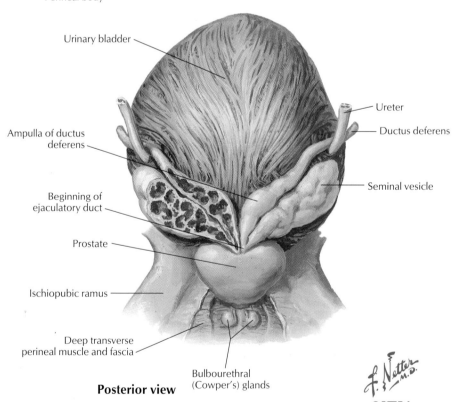

Urinary bladder

Ampulla of ductus deferens

Beginning of ejaculatory duct

Prostate

Ischiopubic ramus

Deep transverse perineal muscle and fascia

Ureter

Ductus deferens

Seminal vesicle

Bulbourethral (Cowper's) glands

Posterior view

The **prostate** is a walnut-sized gland that lies just beneath the urinary bladder and is traversed by the male urethra as it drains urine from the bladder. This gland produces, along with the seminal vesicles, fluid that suspends the sperm and forms the liquid contents of the ejaculated semen. The **seminal vesicles** are tubular glands that lie posterior to the prostate, produce seminal fluid, and join the ductus deferens at the ejaculatory duct.

Scrotum & Testis

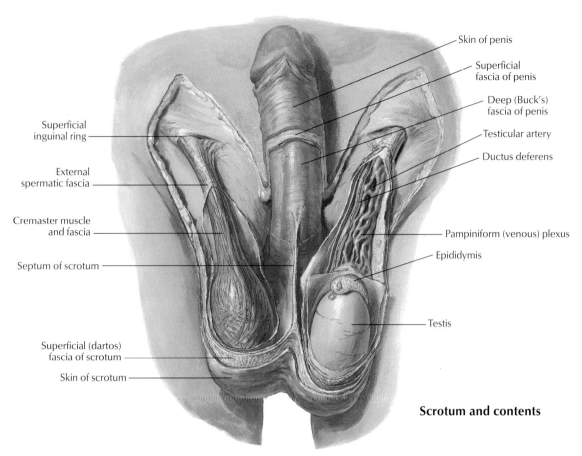

- Skin of penis
- Superficial fascia of penis
- Deep (Buck's) fascia of penis
- Testicular artery
- Ductus deferens
- Pampiniform (venous) plexus
- Epididymis
- Testis

- Superficial inguinal ring
- External spermatic fascia
- Cremaster muscle and fascia
- Septum of scrotum
- Superficial (dartos) fascia of scrotum
- Skin of scrotum

Scrotum and contents

The **scrotum**, the sac of skin behind the penis, holds the testes. The scrotum is covered with pubic hair, to varying degrees depending on the individual.

The **testes**, or **testicles**, are the reproductive glands that produce the male hormone testosterone and the sperm cells. It takes 74 days for sperm cells to develop in the tubules of the testicles. When they emerge, they look mature but lack **motility**, the ability to swim that is crucial for fertility. However, sperm develop motility during the 10 to 12 days they spend traveling through the delivery system. First, the small tubules join to form the **epididymis**, the thin, 20-foot tube coiled behind each testicle. The epididymis then leads to the **vas deferens**, which travels up from the scrotum into the lower pelvis. There, it widens into the **ampulla**, then forms a common ejaculatory duct that travels through the prostate gland into the urethra.

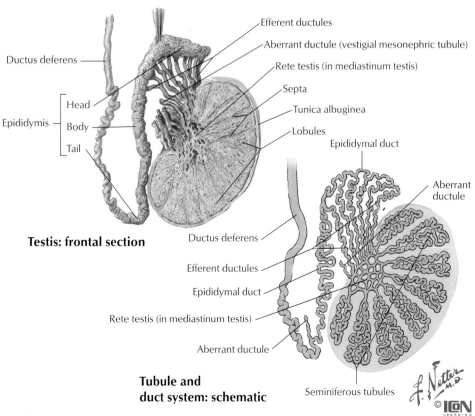

- Ductus deferens
- Epididymis
  - Head
  - Body
  - Tail

- Efferent ductules
- Aberrant ductule (vestigial mesonephric tubule)
- Rete testis (in mediastinum testis)
- Septa
- Tunica albuginea
- Lobules

Testis: frontal section

- Ductus deferens
- Efferent ductules
- Epididymal duct
- Rete testis (in mediastinum testis)
- Aberrant ductule

- Epididymal duct
- Aberrant ductule
- Seminiferous tubules

Tubule and duct system: schematic

155

The Endocrine System

The endocrine system is made up of glands throughout your body that secrete hormones into the blood. These glands include the thyroid, parathyroid, hypothalamus, pineal, pituitary, adrenal, kidney, islands of Langerhans in the pancreas, and testes and ovaries. These glands produce hormones that may affect a specific organ or tissue or the entire body.

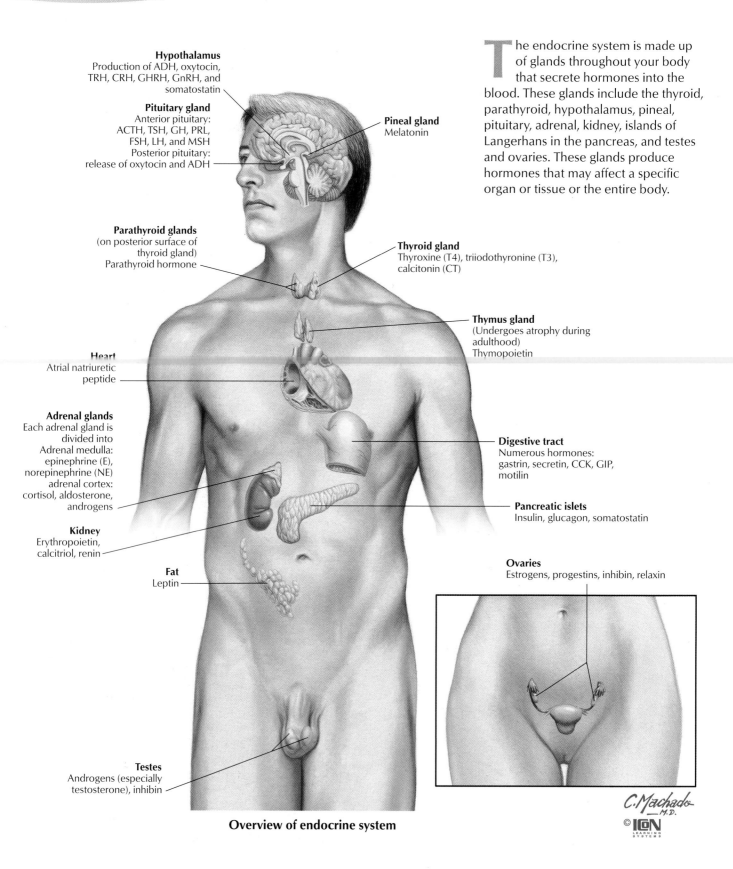

Hypothalamus
Production of ADH, oxytocin, TRH, CRH, GHRH, GnRH, and somatostatin

Pituitary gland
Anterior pituitary:
ACTH, TSH, GH, PRL, FSH, LH, and MSH
Posterior pituitary:
release of oxytocin and ADH

Pineal gland
Melatonin

Parathyroid glands
(on posterior surface of thyroid gland)
Parathyroid hormone

Thyroid gland
Thyroxine (T4), triiodothyronine (T3), calcitonin (CT)

Thymus gland
(Undergoes atrophy during adulthood)
Thymopoietin

Heart
Atrial natriuretic peptide

Adrenal glands
Each adrenal gland is divided into
Adrenal medulla:
epinephrine (E), norepinephrine (NE)
adrenal cortex:
cortisol, aldosterone, androgens

Digestive tract
Numerous hormones:
gastrin, secretin, CCK, GIP, motilin

Pancreatic islets
Insulin, glucagon, somatostatin

Kidney
Erythropoietin, calcitriol, renin

Fat
Leptin

Ovaries
Estrogens, progestins, inhibin, relaxin

Testes
Androgens (especially testosterone), inhibin

Overview of endocrine system

C. Machado M.D.
© ICON LEARNING SYSTEMS

Hypothalamus, Pituitary, & Pineal Glands

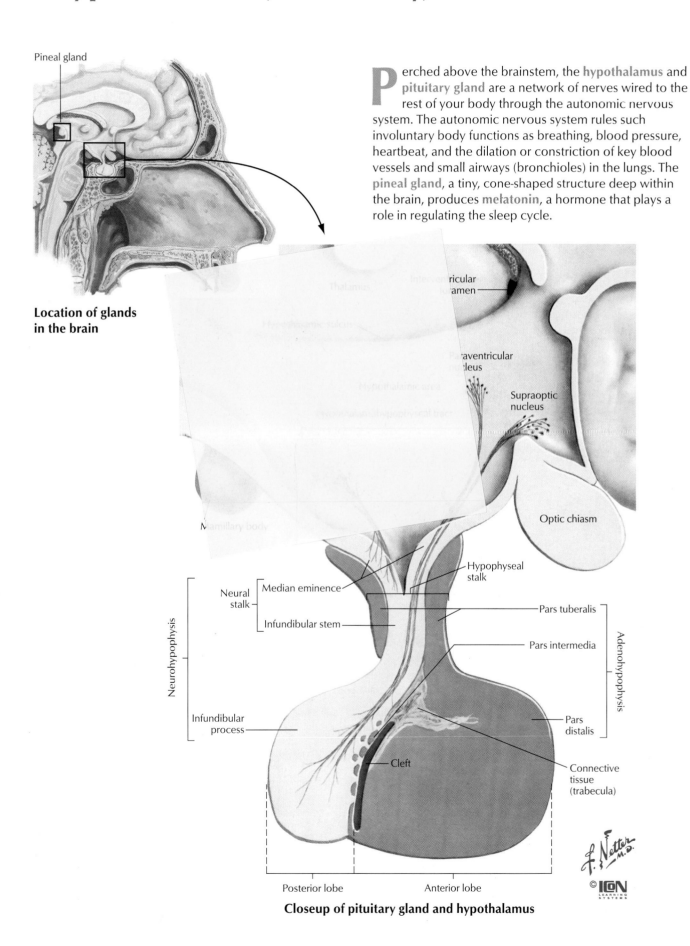

Location of glands in the brain

Perched above the brainstem, the **hypothalamus** and **pituitary gland** are a network of nerves wired to the rest of your body through the autonomic nervous system. The autonomic nervous system rules such involuntary body functions as breathing, blood pressure, heartbeat, and the dilation or constriction of key blood vessels and small airways (bronchioles) in the lungs. The **pineal gland**, a tiny, cone-shaped structure deep within the brain, produces **melatonin**, a hormone that plays a role in regulating the sleep cycle.

Pineal gland

Thalamus

Interventricular foramen

Paraventricular nucleus

Supraoptic nucleus

Optic chiasm

Mamillary body

Hypophyseal stalk

Neural stalk

Median eminence

Pars tuberalis

Infundibular stem

Pars intermedia

Neurohypophysis

Adenohypophysis

Infundibular process

Pars distalis

Cleft

Connective tissue (trabecula)

Posterior lobe

Anterior lobe

Closeup of pituitary gland and hypothalamus

The Endocrine System
Thyroid & Parathyroid Glands

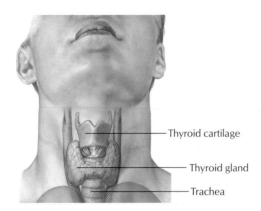

The **thyroid** is a small, butterfly-shaped gland that weighs less than 1 oz. When functioning normally, it perches unobtrusively with its wings wrapped around the front of your windpipe (trachea), below your voice box (larynx). Its slight size could easily fool you into underestimating the thyroid's importance to your health. Yet this gland controls the rate at which every cell, tissue, and organ in your body functions, from your muscles, bones, and skin to your digestive tract, brain, heart, and more.

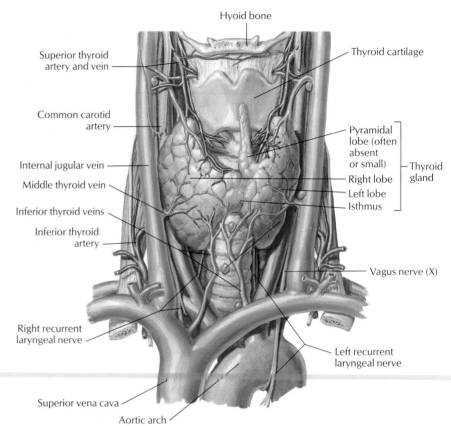

Thyroid gland and surrounding anatomy: anterior view

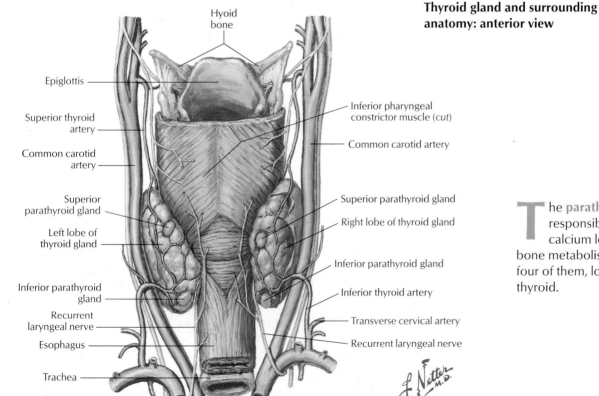

Thyroid and parathyroid glands: posterior view

The **parathyroid glands** are responsible for controlling calcium levels and influencing bone metabolism. There are usually four of them, located behind the thyroid.

Adrenal Glands & Pancreas

The adrenal glands are found above each kidney. These glands secrete many hormones necessary for normal body metabolism and function, including cortisol, which helps the body use sugar and protein, and aldosterone, which maintains your body's balance of salt, potassium, and water.

Inferior phrenic arteries

Inferior vena cava

Esophagus

Left inferior phrenic vein

Right superior suprarenal arteries

Left superior suprarenal arteries

Right suprarenal vein

Left superior suprarenal gland

Right suprarenal gland

Left middle suprarenal artery

Right middle suprarenal artery

Right inferior suprarenal artery

Left inferior suprarenal artery

Left suprarenal vein

Adrenal glands and surrounding anatomy: anterior view

Inferior vena cava

Abdominal aorta

The pancreas is both an exocrine gland, which produces digestive enzymes, and an endocrine gland that consists of small collections of cells called the islets of Langerhans. Some cells in the islets produce the hormone glucagon, a fuel mobilization hormone that acts on glycogen stores in the liver or directly on fat. Other islets cells produce somatostatin that acts on the digestive tract. The islets also produce insulin, a fuel storage hormone that helps cells in the body take up and store glucose.

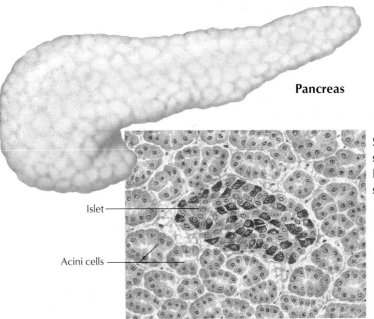

Pancreas

Selection of pancreas showing an islet of Langerhaus cells surrounded by acini

Islet

Acini cells

The Endocrine System
Menstrual Cycle

The **menstrual cycle** (ideally shown as a 28-day cycle) is divided into three phases: the **follicular phase** (development of a mature follicle or egg), the **ovulatory phase** (the egg is ovulated or released into the reproductive tract), and the **luteal phase** (preparation of the uterus for implantation if the egg is fertilized by a sperm). If fertilization and implantation do not occur, the **uterine lining** is sloughed off and menstruation begins. Each phase is carefully regulated by the levels of various hormones.

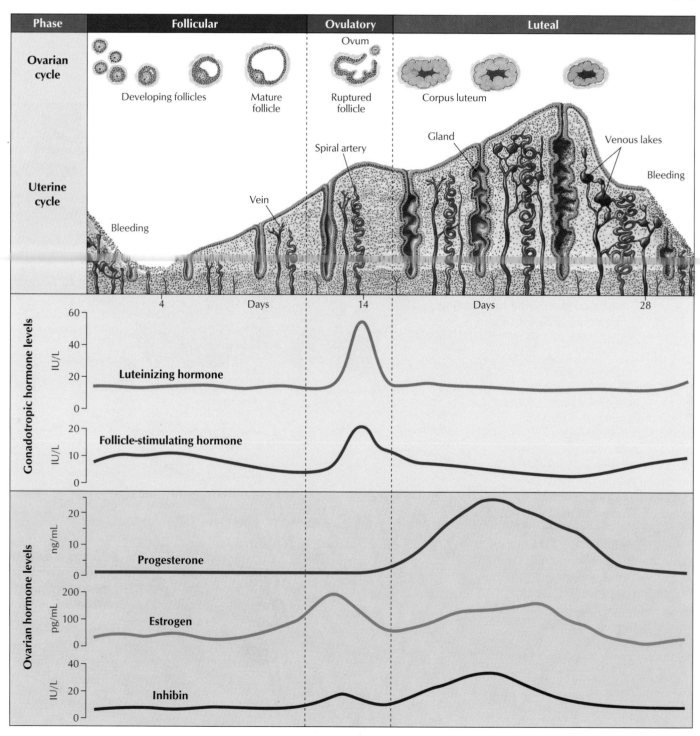

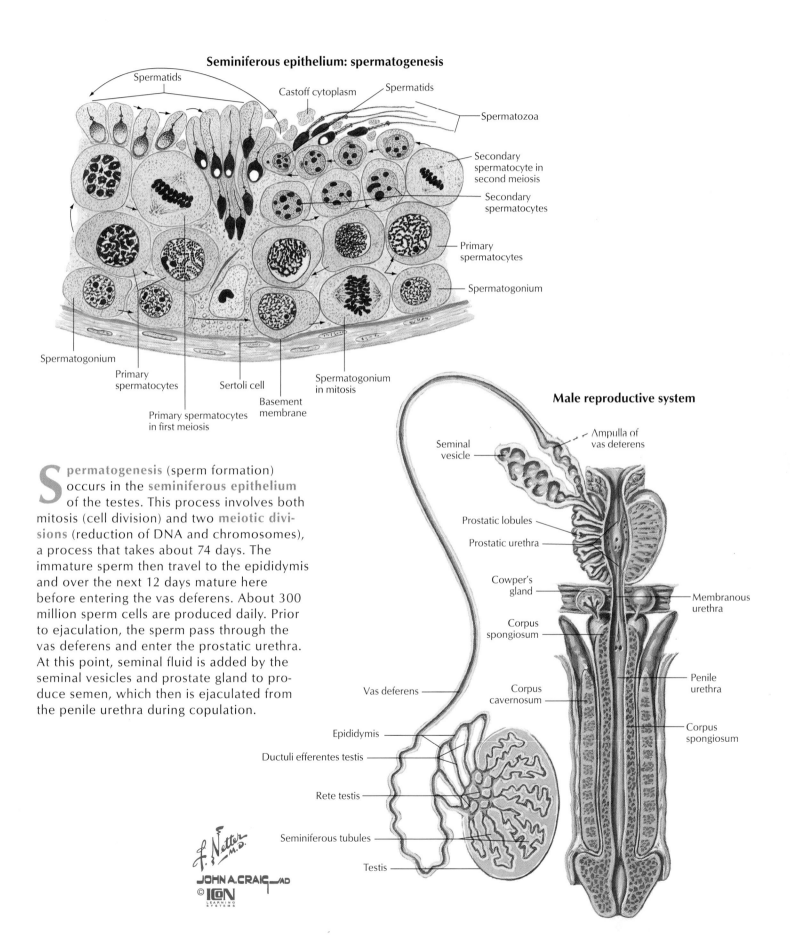

Seminiferous epithelium: spermatogenesis

Spermatids

Castoff cytoplasm

Spermatids

Spermatozoa

Secondary spermatocyte in second meiosis

Secondary spermatocytes

Primary spermatocytes

Spermatogonium

Spermatogonium

Primary spermatocytes

Sertoli cell

Spermatogonium in mitosis

Basement membrane

Primary spermatocytes in first meiosis

Male reproductive system

Seminal vesicle

Ampulla of vas deferens

Prostatic lobules

Prostatic urethra

Cowper's gland

Membranous urethra

Corpus spongiosum

Corpus cavernosum

Penile urethra

Corpus spongiosum

Vas deferens

Epididymis

Ductuli efferentes testis

Rete testis

Seminiferous tubules

Testis

Spermatogenesis (sperm formation) occurs in the **seminiferous epithelium** of the testes. This process involves both mitosis (cell division) and two **meiotic divisions** (reduction of DNA and chromosomes), a process that takes about 74 days. The immature sperm then travel to the epididymis and over the next 12 days mature here before entering the vas deferens. About 300 million sperm cells are produced daily. Prior to ejaculation, the sperm pass through the vas deferens and enter the prostatic urethra. At this point, seminal fluid is added by the seminal vesicles and prostate gland to produce semen, which then is ejaculated from the penile urethra during copulation.

JOHN A.CRAIG—AD

© ICN LEARNING SYSTEMS

References

The following were additional resources for text in the book:

Cardiovascular System
P. 83: Text adapted from Aetna Intellihealth.com, featuring Harvard Medical School's consumer health information (http://www.intelihealth.com).

P 84: Text excerpted from Hansen, John. *Netter's Clinical Anatomy*. Carlstadt, NJ: ICON Learning Systems, 2005, p. 340.

Blood and Lymphatic System

P 98: Text adapted from Aetna Intellihealth.com, featuring Harvard Medical School's consumer health information (http://www.intelihealth.com).

P 100: Text adapted from the National Institute of Allergy and Infectious Diseases, NIAID NetNews, *The Immune System* (http://www.niaid.nih.gov/final/immun/immun.htm).

P 101: Text adapted from Aetna Intellihealth.com, featuring Harvard Medical School's consumer health information (http://www.intelihealth.com).

Nervous system:
P 103: Text adapted from the National Center for Biotechnology Information, National Institutes of Health, *Genes and Disease: The Nervous System* (http://www.ncbi.nlm.nih.gov/).

Urinary System
P 138: Text adapted from National Urologic and Kidney Diseases Information Clearinghouse, National Institutes of Health, *Your Kidneys and How They Work* (http://kidney.niddk.nih.gov/index.htm).

Endocrine System
P 156: Text adapted from U.S. National Cancer Institute's Surveillance, Epidemiology and End Results (SEER) Program Training web site, National Institutes of Health, *Endocrine Glands and Their Hormones* (http://healthservices.cancer.gov/seer-medicare).

P 159: Text adapted from Patient Information Publications, NIH Clinical Center, National Institutes of Health, *Managing Adrenal Insufficiency* (http://clinicalcenter.nih.gov/ccc/patient_education/pepubs/mngadrins.pdf).

Frank H. Netter, MD

Frank H. Netter was born in 1906 in New York City. He studied art at the Art Student's League and the National Academy of Design before entering medical school at New York University, where he received his MD degree in 1931. During his student years, Dr. Netter's notebook sketches attracted the attention of the medical faculty and other physicians, allowing him to augment his income by illustrating articles and textbooks. He continued illustrating as a sideline after establishing a surgical practice in 1933, but he ultimately opted to give up his practice in favor of a full-time commitment to art. After service in the United States Army during World War II, Dr. Netter began his long collaboration with the CIBA Pharmaceutical Company (now Novartis Pharmaceuticals). This 45-year partnership resulted in the production of the extraordinary collection of medical art so familiar to physicians and other medical professionals worldwide.

Icon Learning Systems acquired the Netter Collection in July 2000 and continues to update Dr. Netter's original paintings and to add newly commissioned paintings by artists trained in the style of Dr. Netter.

Dr. Netter's works are among the finest examples of the use of illustration in the teaching of medical concepts. The 13-book *Netter Collection of Medical Illustrations*, which includes the greater part of the more than 20,000 paintings created by Dr. Netter, became and remains one of the most famous medical works ever published. *The Netter Atlas of Human Anatomy*, first published in 1989, presents the anatomical paintings from the Netter Collection. Now translated into 11 languages, it is the anatomy atlas of choice among medical and health professions students the world over.

The Netter illustrations are appreciated not only for their aesthetic qualities but, more important, for their intellectual content. As Dr. Netter wrote in 1949 "... clarification of a subject is the aim and goal of illustration. No matter how beautifully painted, how delicately and subtly rendered a subject may be, it is of little value as a *medical illustration* if it does not serve to make clear some medical point." Dr. Netter's planning, conception, point of view, and approach are what inform his paintings and what makes them so intellectually valuable.

Frank H. Netter, MD, physician and artist, died in 1991.